PRAISE FOR ANDREW HINTON

From "AI Basics for Managers:"

"a beacon, guiding non-tech leaders through the often murky waters of artificial intelligence"

— KIRAN

"this book is your ticket to understanding AI's potential and implementing it effectively"

— KAM

"the book was written in an easy-to-understand style, with enough info to make it interesting and informational, but not too much technical data ... the futuristic images, nice touch and beautifully reflects the topic of the book"

— SHAMSA K

"book's holistic approach, ranging from AI's history to its potential impact on contemporary business, is truly commendable"

— AES

ESSENTIAL MATH FOR AI

ESSENTIAL MATH FOR AI

EXPLORING LINEAR ALGEBRA, PROBABILITY
AND STATISTICS, CALCULUS, GRAPH THEORY,
DISCRETE MATHEMATICS, NUMERICAL
METHODS, OPTIMIZATION TECHNIQUES,
AND MORE

AI FUNDAMENTALS

ANDREW HINTON

B

Book Bound
STUDIOS

To the pioneers of the past and the innovators of the future, whose curiosity and determination continue to bridge the realms of mathematics and artificial intelligence. This book is dedicated to you.

Mathematics is the door and key to the sciences.

— ROGER BACON

CONTENTS

Receive Your Free Copy of The Power of AI

SCAN ME

Or visit:
bookboundstudios.wixsite.com/andrew-hinton

THE ROLE OF MATHEMATICS IN ARTIFICIAL INTELLIGENCE

In the rapidly evolving world of artificial intelligence (AI), one might be tempted to think that the key to success lies solely in mastering the latest programming languages, algorithms, and software tools. While these skills are undoubtedly necessary, there is a foundational element that underpins the entire field of AI: mathematics. The inextricable link between mathematics and AI is a testament to the fact that, at its core, AI is a discipline rooted in mathematical principles and concepts. This connection is not a mere coincidence but rather a reflection of mathematics's fundamental role in shaping the development and application of AI technologies.

Mathematics provides the language and framework to understand, analyze, and manipulate the complex data structures and algorithms that drive AI systems. From the basic building blocks of linear algebra and calculus to the more advanced realms of probability theory, optimization, and graph theory, mathematics is the backbone of AI, enabling researchers and practitioners to push the boundaries of what is possible in this exciting field.

As AI continues to permeate every aspect of our lives, from self-driving cars and virtual personal assistants to advanced medical diagnostics and financial forecasting, understanding the underlying mathematical principles cannot be overstated. By delving into the essential math for AI, we not only equip ourselves with the tools necessary to navigate this complex landscape but also gain a deeper appreciation for the intricate interplay between mathematics and AI that has given rise to some of the most groundbreaking innovations of our time.

In this book, we will embark on a journey to explore the fascinating world of AI mathematics, shedding light on the key concepts and techniques that form the foundation of this rapidly growing field. Through a unique blend of theory and practical applications, we will empower AI enthusiasts with the essential mathematical knowledge needed to excel in their pursuits and contribute to the ongoing development of AI technologies.

As we prepare to delve into the multifaceted world of AI mathematics, let us remember that the challenges we face today are merely stepping stones to the incredible innovations that await us. By embracing the power of mathematics and harnessing its potential, we can unlock the door to a new era of AI-driven advancements that will undoubtedly reshape the world as we know it.

A Brief History of Mathematical Foundations in AI

The story of artificial intelligence (AI) is deeply intertwined with the history of mathematics. From the earliest days of AI research, mathematical concepts and techniques have shaped the field's development. In this section, we will briefly journey through the annals of AI history, highlighting the key mathematical milestones that have contributed to the field's growth and evolution.

The roots of AI can be traced back to the ancient world, where philosophers and mathematicians like Aristotle and Euclid laid the groundwork for logic and geometry. These early thinkers grappled with questions about the nature of intelligence and the possibility of creating machines that could think and reason like humans. Their ideas would later inspire the pioneers of AI, who sought to build on these foundations and develop mathematical intelligence models.

In the 20th century, AI began to take shape, driven by the work of mathematicians and computer scientists such as Alan Turing, John von Neumann, and Claude Shannon. Turing's groundbreaking work on computability theory and the Turing machine provided a formal framework for understanding the limits of computation, while von Neumann's contributions to game theory and cellular automata laid the groundwork for the study of complex systems and decision-making in AI. Shannon's information theory, meanwhile, offered a mathematical basis for understanding the transmission and processing of information, which would become central to AI research.

As AI research progressed, new mathematical tools and techniques were developed to tackle the challenges of modeling and simulating intelligent behavior. Probability theory and statistics emerged as essential tools for dealing with uncertainty and making inferences from data, while optimization techniques provided a means of finding the best solutions to complex problems. Graph theory and combinatorics played a crucial role in understanding the structure and properties of networks, which are fundamental to many AI applications.

In the latter half of the 20th century, the advent of machine learning and neural networks brought new mathematical challenges to the forefront of AI research. Researchers drew on concepts from linear algebra, calculus, and numerical analysis to develop algorithms that could learn from data and adapt their behavior over time. These advances paved the way for developing robust AI systems tackling various tasks, from image recognition to natural language processing.

Today, the field of AI continues to evolve at a rapid pace, driven by ongoing advances in mathematics and computer science. As we explore the essential math for AI in this book, we will delve into the rich tapestry of mathematical ideas that underpin the field, from the foundational concepts of logic and probability to the cutting-edge techniques of deep learning and reinforcement learning. By understanding the mathematical foundations of AI, we can better appreciate the intricate interplay between theory and practice that defines this exciting and ever-changing discipline.

Empowering AI Enthusiasts with Essential Mathematical Knowledge

The primary purpose of this book is to empower AI enthusiasts, be they students, researchers, or professionals, with the essential mathematical knowledge required to excel in the field of artificial intelligence. As the world increasingly relies on AI-driven technologies, it is crucial for those involved in developing and implementing these systems to possess a strong foundation in the mathematical concepts underpinning their functionality.

This book aims to provide a comprehensive yet accessible guide to the mathematical principles that form the backbone of AI. By presenting these concepts clearly and concisely, we aim to make the subject matter approachable for readers of all levels of expertise, from those just beginning their journey into AI to seasoned professionals seeking to deepen their understanding.

Throughout the chapters, we will explore a wide range of topics, including linear algebra, probability theory, optimization, and more, all of which play a crucial role in designing and operating AI systems. By providing a thorough grounding in these areas, we aim to equip our readers with the tools they need to tackle the complex challenges that arise in the development of AI technologies.

Moreover, this book is designed to balance theory and practical applications. While it is essential to understand the underlying mathematical principles, it is equally important to see how these concepts can be applied in real-world situations. To this end, we will include numerous examples and case studies that demonstrate the relevance of the mathematical concepts discussed in the context of AI.

In conclusion, this book aims to empower AI enthusiasts with the essential mathematical knowledge needed to excel in the field of artificial intelligence. By providing a comprehensive and accessible guide to the mathematical foundations of AI, we hope to prepare our readers for the challenges and opportunities in the exciting world of AI innovation.

Preparing for the Mathematical Challenges of Tomorrow's AI Innovations

As we stand on the precipice of a new era in artificial intelligence, it is crucial to recognize mathematics's indispensable role in shaping the future of this rapidly evolving field. The innovations of tomorrow will undoubtedly present a myriad of mathematical challenges, and it is our responsibility as AI enthusiasts, researchers, and practitioners to be well-equipped with the essential mathematical knowledge to tackle these obstacles head-on.

Throughout this book, we have embarked on a journey to explore the multifaceted world of AI mathematics, delving into the theoretical underpinnings and practical applications that form the backbone of artificial intelligence. By providing a comprehensive understanding of the mathematical foundations that govern AI, we aim to empower our readers with the tools and confidence necessary to contribute to this exciting field's ongoing development and advancement.

As we look ahead to the future of artificial intelligence, it is important to remember that the innovations of tomorrow will be built upon the mathematical foundations of today. By fostering a deep understanding of the essential mathematical concepts underlying AI, we can ensure that we are prepared to navigate this dynamic field's complex and ever-changing landscape.

In conclusion, the role of mathematics in artificial intelligence cannot be overstated. As we continue to push the boundaries of what is possible with AI, we must remain steadfast in our commitment to mastering the mathematical principles that drive these innovations. By doing so, we can unlock the full potential of artificial intelligence and usher in a new era of technological breakthroughs that will undoubtedly reshape the world as we know it.

So, as you turn the pages of this book and immerse yourself in the fascinating world of AI mathematics, remember that you are not only gaining valuable knowledge but also preparing yourself for the mathematical challenges that lie ahead in the ever-evolving realm of artificial intelligence. Embrace the journey, and let the power of mathematics guide you toward a brighter, AI-driven future.

1

LINEAR ALGEBRA: THE FOUNDATION OF MACHINE LEARNING

I n the world of artificial intelligence (AI), one might wonder what role mathematics, specifically linear algebra, plays in developing and advancing this cutting-edge technology. As it turns out, linear algebra is not only a crucial component of AI but also the foundation upon which machine learning algorithms are built. In this chapter, we will delve into the fascinating world of linear algebra and explore its significance in AI, providing you with the essential knowledge to understand and master this critical mathematical discipline.

Linear algebra, at its core, is the study of vector spaces and the linear relationships between them. It is a branch of mathematics that deals with vectors, matrices, and systems of linear equations, all fundamental concepts in AI. As we progress through this chapter, we will uncover the reasons behind the importance of linear algebra in AI and how it enables machines to learn, adapt, and make intelligent decisions.

One of the primary reasons linear algebra is essential to AI is its ability to represent and manipulate large amounts of data efficiently. In the world of AI, data is king. The more data an algorithm can process, the better it can learn and make accurate predictions. Linear algebra provides the tools necessary to handle vast quantities of data in the form of vectors and matrices, which are the building blocks of this mathematical field. By understanding how to work with these structures, we can create powerful machine-learning algorithms that can process and analyze data at an incredible scale.

Another reason linear algebra is vital to AI is its role in transforming and manipulating data. Machine learning algorithms often require data to be transformed or manipulated in specific ways to extract valuable insights and patterns. Linear algebra provides a set of operations and transformations, such as matrix multiplication and inversion, that allow us to modify data in a structured and efficient manner. By mastering these techniques, we can preprocess and transform data to be more suitable for machine learning algorithms, ultimately improving their performance and accuracy.

Eigenvalues and eigenvectors, two essential concepts in linear algebra, also play a significant role in AI. They uncover hidden patterns and structures within data, which can be invaluable in machine learning applications. By understanding the principles behind eigenvalues and eigenvectors, we can develop algorithms to identify and exploit these patterns, leading to more effective and intelligent AI systems.

In conclusion, linear algebra is the foundation for building machine learning and AI. By understanding and mastering the concepts of vectors, matrices, and the various operations and transformations associated with them, we can develop powerful AI algorithms capable of learning from vast amounts of data and making intelligent decisions. As we progress through this chapter, we will explore these concepts in greater detail, providing you with the essential mathematical knowledge to excel in AI.

Understanding Vectors and Matrices: The Building Blocks of Linear Algebra

This section will delve into the fascinating world of vectors and matrices, the fundamental building blocks of linear algebra. These mathematical entities play a crucial role in developing and implementing artificial intelligence (AI) and machine learning algorithms. By the end of this section, you will have a solid understanding of vectors and matrices and be well-equipped to appreciate their significance in the realm of AI.

Vectors: The Backbone of Data Representation

A vector is a mathematical object that represents both magnitude and direction. In the context of AI, vectors are used to represent data points in a multi-dimensional space. For instance, consider a simple example of a movie rating system. Each movie can be represented by a vector containing ratings in various categories, such as action, romance, and comedy. By representing data in this manner, we can easily compare and analyze different movies based on their ratings.

Vectors are typically represented as an ordered list of numbers enclosed in square brackets. For example, a 3-dimensional vector can be represented as [x, y, z], where x, y, and z are the magnitudes of the vector in the respective dimensions. The number of dimensions in a vector is called its size or dimensionality.

Matrices: Organizing Data for Efficient Processing

A matrix is a rectangular array of numbers, symbols, or expressions arranged in rows and columns. In AI, matrices organize and manipulate large amounts of data efficiently. They are instrumental in representing and processing data through images, graphs, and networks.

A matrix is a collection of vectors where each row or column represents a vector. The size of a matrix is defined by the number of rows (m) and columns (n) and is denoted as an m × n matrix. For example, a 3 × 2 matrix has 3 rows and 2 columns.

The Connection Between Vectors and Matrices

Vectors and matrices are intrinsically connected, as a matrix can be considered a collection of vectors. This connection is vital in AI, as it allows us to manipulate and transform data in a structured and efficient manner. We can perform complex operations and transformations essential for machine learning algorithms by representing data as vectors and matrices.

In the following sections, we will explore various matrix operations and transformations, as well as the concepts of eigenvalues and eigenvectors, which play a pivotal role in uncovering hidden patterns in data. By mastering these concepts, you will be well on your way to building a solid foundation in linear algebra, which is essential for success in AI.

Matrix Operations and Transformations: Manipulating Data for Machine Learning

In the fascinating world of artificial intelligence, linear algebra plays a pivotal role in shaping the foundation of machine learning algorithms. As we delve deeper into this realm, we come across matrix operations and transformations, which are crucial in manipulating data for machine learning. In this section, we will explore the various matrix operations, their significance, and how they contribute to the development of AI.

Matrix operations are mathematical procedures that involve the manipulation of matrices, which are essentially rectangular arrays of numbers, symbols, or expressions. These operations are vital in machine learning, enabling us to efficiently perform complex calculations and transformations on large datasets. Let's examine some of the most common matrix operations and their significance in AI.

- **Matrix Addition and Subtraction:** These are the most basic operations that involve adding or subtracting corresponding elements of two matrices. This operation is particularly useful in machine learning when we combine or compare datasets, update weights in neural networks, or calculate the error between predicted and actual values.
- **Matrix Multiplication:** This operation involves multiplying two matrices in a specific order, resulting in a new matrix. Matrix multiplication is essential in machine learning, as it allows us to apply transformations to our data, such as scaling, rotation, and translation. Furthermore, it significantly implements various algorithms, such as linear regression and neural networks.
- **Matrix Transpose:** The transpose of a matrix is obtained by interchanging its rows and columns. This operation is particularly useful in machine learning when we need to rearrange data for specific calculations or when working with algorithms requiring input data to be in a particular format.
- **Matrix Inversion:** The inverse of a square matrix is a matrix that, when multiplied with the original matrix, results in the identity matrix. Matrix inversion is a crucial operation in machine learning, as it helps solve systems of linear equations, which are often encountered in various algorithms, such as linear regression and support vector machines.

Now that we have a basic understanding of matrix operations let's discuss matrix transformations. In machine learning, transformations are essential in preprocessing data, reducing dimensions, and extracting features. Some common matrix transformations include:

- **Scaling:** This transformation involves multiplying the elements of a matrix by a scalar value, effectively changing the size of the data. Scaling is often used in machine learning to normalize data, ensuring that all features have the same range of values, which helps improve the performance of algorithms.
- **Rotation:** This transformation involves rotating the data in a multi-dimensional space, which can be achieved by multiplying the data matrix with a rotation matrix. Rotation is useful in machine learning for data visualization, dimensionality reduction, and improving the performance of certain algorithms.
- **Translation:** This transformation involves adding or subtracting a constant value to the elements of a matrix, effectively shifting the data in a multi-dimensional space. Translation is often used in machine learning to center the data around the origin, which can improve the performance of various algorithms.

In conclusion, matrix operations and transformations are indispensable tools in machine learning. They enable us to manipulate and preprocess data efficiently, paving the way for developing robust AI algorithms. Mastering these concepts can build a strong foundation in AI and contribute to the ever-evolving landscape of artificial intelligence.

Eigenvalues and Eigenvectors: Uncovering Hidden Patterns in Data

In artificial intelligence, the ability to identify patterns and extract valuable information from vast amounts of data is crucial. One of the most powerful tools for achieving this is the concept of eigenvalues and eigenvectors. In this section, we will delve into the world of these mathematical entities, exploring their significance and applications in machine learning.

The Essence of Eigenvalues and Eigenvectors

Before we dive into the intricacies of eigenvalues and eigenvectors, let's first establish a fundamental understanding of these terms. In linear algebra, an eigenvector is a non-zero vector that, when multiplied by a matrix, results in a scalar multiple of itself. This scalar multiple is known as the eigenvalue. Mathematically, this relationship can be expressed as:

$$A * v = \lambda * v$$

Here, A represents the matrix, v is the eigenvector, and λ is the eigenvalue.

The Intuition Behind Eigenvalues and Eigenvectors

Let's consider a real-world analogy to grasp the concept of eigenvalues and eigenvectors better. Imagine a group of people standing on a platform that can rotate around a central axis. Each person's position changes as the platform rotates, but the central axis remains fixed. In this scenario, the central axis represents the eigenvector, and the rotation angle corresponds to the eigenvalue.

In the context of data analysis, eigenvalues, and eigenvectors can be thought of as the "central axes" around which the data points are organized. By identifying these axes, we can uncover hidden patterns and relationships within the data, enabling us to make more informed decisions and predictions.

Applications of Eigenvalues and Eigenvectors in Machine Learning

Eigenvalues and eigenvectors are pivotal in various machine

learning algorithms, particularly in dimensionality reduction and data compression. Some notable applications include:

Principal Component Analysis (PCA): PCA is a widely-used technique for reducing the dimensionality of large datasets while preserving as much information as possible. By calculating the eigenvectors and eigenvalues of the data's covariance matrix, PCA identifies the directions (principal components) along which the variance of the data is maximized.

Singular Value Decomposition (SVD): SVD is a matrix factorization technique that decomposes a given matrix into three matrices, one containing the eigenvectors. SVD is used in numerous machine learning applications, such as image compression, natural language processing, and recommender systems.

Spectral Clustering: In spectral clustering, eigenvalues and eigenvectors are employed to partition data points into distinct clusters based on similarity. This technique is handy for identifying non-linear structures within the data.

Mastering Eigenvalues and Eigenvectors for a Strong AI Foundation

In conclusion, eigenvalues and eigenvectors are indispensable tools in artificial intelligence and machine learning. Understanding their underlying principles and applications, you will be better equipped to tackle complex data analysis tasks and develop more effective machine learning models. As you continue to explore the fascinating world of linear algebra, remember that the key to success lies in building a solid foundation and continually expanding your knowledge.

Applications of Linear Algebra in Machine Learning Algorithms

In this section, we will delve into the fascinating world of machine learning and explore how linear algebra plays a crucial role in developing and optimizing various algorithms. By understanding the applications of linear algebra in machine learning, we can appreciate its significance and harness its power to create more efficient and accurate models.

Image Recognition and Computer Vision

One of the most popular applications of machine learning is image recognition, where computers are trained to identify and classify objects within images. Linear algebra is at the heart of this process, as images can be represented as matrices of pixel values. By applying matrix operations and transformations, we can manipulate these images, extract features, and reduce dimensions for more efficient processing. Furthermore, eigenvectors and eigenvalues can be used to identify principal components essential in techniques like Principal Component Analysis (PCA) for dimensionality reduction and data compression.

Natural Language Processing

Linear algebra is also a key component in natural language processing (NLP), which involves computer computers' analysis and generation of human language. In NLP, text data is often represented as vectors in high-dimensional spaces, where each dimension corresponds to a unique word or feature. By applying linear algebra techniques, we can measure the similarity between texts, perform sentiment analysis, and even generate new text based on existing patterns. Techniques such as Word2Vec and Latent Semantic Analysis (LSA) rely heavily on linear algebra concepts to create meaningful representations of text data.

Recommender Systems

Recommender systems are widely used in e-commerce, content platforms, and social media to suggest products, articles, or connections based on user preferences and behavior. Linear algebra plays a vital role in developing these systems, as it allows us to represent user preferences and item features as vectors and matrices. By applying matrix factorization techniques, we can uncover latent factors that explain observed user-item interactions and make personalized recommendations. Singular Value Decomposition (SVD) and collaborative filtering are popular linear algebra-based techniques used in recommender systems.

Deep Learning and Neural Networks

Deep learning, a subset of machine learning, involves using artificial neural networks to model complex patterns and make predictions. Linear algebra is fundamental to the structure and functioning of these networks, as they consist of layers of interconnected nodes, each performing linear transformations on input data. By adjusting the weights and biases of these connections through backpropagation, the network can learn to make accurate predictions. Convolutional Neural Networks (CNNs), a popular type of deep learning architecture for image recognition, rely heavily on linear algebra operations such as convolution and pooling to process and analyze input data.

In conclusion, linear algebra is a powerful and versatile tool that underpins many machine learning algorithms. By mastering the concepts and techniques of linear algebra, we can build a strong foundation for understanding and developing advanced AI models. As AI continues to evolve, linear algebra will undoubtedly remain essential in the quest for more intelligent and efficient machines.

Mastering Linear Algebra for a Strong AI Foundation

In conclusion, linear algebra plays a pivotal role in artificial intelligence. As we have explored throughout this chapter, the concepts and techniques of linear algebra are deeply intertwined with the foundations of machine learning, providing the necessary tools to manipulate, analyze, and interpret vast amounts of data. By mastering linear algebra, one can unlock the true potential of AI and contribute to the development of innovative solutions that can revolutionize various aspects of our lives.

The journey through linear algebra began with understanding the basic building blocks: vectors and matrices. These fundamental elements serve as the backbone of linear algebra, allowing us to represent complex data structures and perform intricate calculations. As we delved deeper into the subject, we discovered the power of matrix operations and transformations, which enable us to manipulate data in ways that facilitate machine learning processes.

One of the most intriguing aspects of linear algebra is the concept of eigenvalues and eigenvectors. These unique properties of matrices allow us to uncover hidden patterns and structures within data, providing valuable insights that machine learning algorithms can harness. By understanding the significance of eigenvalues and eigenvectors, we can develop more efficient and accurate models that tackle various AI challenges.

Throughout this chapter, we have also highlighted various linear algebra applications in machine learning algorithms. From linear regression to neural networks, the influence of linear algebra is omnipresent, showcasing its importance in the field of AI. By mastering these techniques, aspiring AI professionals can equip themselves with the knowledge and skills to excel in this rapidly evolving domain.

In summary, linear algebra is an indispensable component of artificial intelligence, serving as the foundation upon which machine learning algorithms are built. Mastering the concepts and techniques presented in this chapter can establish a strong foundation in AI and contribute to advancing this transformative technology. As we continue to explore the vast landscape of artificial intelligence, let us remember the significance of linear algebra and its power in shaping AI's future.

Chapter Summary

- Linear algebra is the foundation of artificial intelligence and machine learning, providing the necessary tools to manipulate, analyze, and interpret vast amounts of data.
- Vectors and matrices are the fundamental building blocks of linear algebra, allowing for the representation of complex data structures and the performance of intricate calculations.
- Matrix operations and transformations, such as addition, subtraction, multiplication, inversion, and transpose, are crucial for manipulating data to facilitate machine learning processes.
- Eigenvalues and eigenvectors are unique properties of matrices that uncover hidden patterns and structures within data, providing valuable insights for machine learning algorithms.
- Linear algebra plays a significant role in various machine learning applications, including image recognition, natural language processing, recommender systems, and deep learning.
- Principal Component Analysis (PCA) and Singular Value Decomposition (SVD) are popular linear algebra-based techniques used for dimensionality reduction and data compression in machine learning.
- Convolutional Neural Networks (CNNs), a popular deep learning architecture for image recognition, rely heavily on linear algebra operations such as convolution and pooling to process and analyze input data.
- Mastering linear algebra concepts and techniques is essential for building a solid foundation in artificial intelligence and contributing to developing innovative AI solutions.

2

PROBABILITY AND STATISTICS: UNDERSTANDING DATA AND UNCERTAINTY

W hen working with artificial intelligence (AI), making informed decisions and predictions based on data is crucial. As AI systems continue to permeate various aspects of our lives, from healthcare to finance, understanding the underlying mathematical concepts becomes increasingly important. One such area of mathematics that plays a pivotal role in AI is probability and statistics. This chapter will delve into the fascinating world of probability and statistics, exploring their significance in AI and how they help us make sense of the vast amounts of data we encounter daily.

Probability and statistics are two interconnected branches of mathematics that deal with the analysis and interpretation of data. While probability theory focuses on the likelihood of events occurring, statistics is concerned with data collection, organization, analysis, interpretation, and presentation. Together, they provide a powerful toolkit for understanding and managing uncertainty, a fundamental aspect of AI.

In AI, probability and statistics are used to model and predict various phenomena, such as the behavior of users, the performance of algorithms, and the outcomes of experiments. By quantifying uncertainty and extracting meaningful insights from data, AI developers can create more robust and reliable systems that can adapt to new information and make better decisions.

The chapter will begin with an overview of the fundamentals of probability theory, including the basic concepts and principles that underpin this fascinating area of mathematics. We will then move on to descriptive statistics, where we will learn how to summarize and visualize data meaningfully. This will be followed by a discussion of inferential statistics, which involves drawing conclusions from data and predicting future events.

Next, we will explore the concept of Bayesian inference, a powerful statistical technique that allows us to update our beliefs about the world as new data becomes available. This approach is particularly relevant in AI, enabling machines to learn from experience and improve their performance over time.

Finally, we will conclude the chapter by reflecting on the role of probability and statistics in AI development, highlighting their importance in building intelligent systems that can navigate the complexities of the real world.

As we embark on this journey through the realm of probability and statistics, we invite you to join us in discovering the beauty and power of these mathematical tools and how they can help us unlock the full potential of AI. With a blend of expository, descriptive, narrative, and persuasive writing, this chapter aims to provide a comprehensive and engaging introduction to the essential math for AI. So, let's dive in and explore the fascinating world of probability and statistics in AI!

Fundamentals of Probability Theory

In artificial intelligence (AI), making informed data-based decisions is crucial. Probability theory is the backbone of this decision-making process, providing a mathematical framework to quantify uncertainty and make predictions. In this section, we will delve into the fundamentals of probability theory, exploring its core concepts and their applications in AI.

Basic Concepts of Probability

- **Sample Space and Events:** The sample space (S) represents the set of all possible outcomes of a random experiment. An event is a subset of the sample space of one or more outcomes. For example, when rolling a six-sided die, the sample space is S = {1, 2, 3, 4, 5, 6}, and an event could be rolling an even number, represented as E = {2, 4, 6}.
- **Probability Measure:** A probability measure (denoted as P) is a function that assigns a value between 0 and 1 to each event, indicating the likelihood of that event occurring. The sum of probabilities for all events in the sample space must equal 1. In our die-rolling example, the probability of rolling an even number is P(E) = 3/6 = 1/2.
- **Conditional Probability:** Conditional probability (P(A|B)) represents the probability of event A occurring, given that event B has occurred. This concept is essential in AI, allowing us to update our beliefs based on new information. For instance, if we know that a rolled die shows an even number, the probability of it being a 4 is P(4|even) = 1/3.

Probability Rules and Theorems

- **Addition Rule:** The addition rule states that the probability of either event A or event B occurring equals the sum of their individual probabilities minus the probability of both events occurring simultaneously. Mathematically, $P(A \cup B) = P(A) + P(B) - P(A \cap B)$.
- **Multiplication Rule:** The multiplication rule is used to calculate the probability of two events occurring simultaneously. If events A and B are independent (i.e., the occurrence of one does not affect the other), then $P(A \cap B) = P(A) \times P(B)$. If they are dependent, $P(A \cap B) = P(A) \times P(B|A)$.
- **Bayes' Theorem:** Bayes' theorem is a cornerstone of probability theory, allowing us to update our beliefs based on new evidence. It states that $P(A|B) = P(B|A) \times P(A) / P(B)$. This theorem is often used in AI to refine predictions as more data becomes available.

Applications of Probability Theory in AI

- **Decision-making:** AI systems often need to make decisions based on uncertain information. Probability theory provides a framework for weighing the likelihood of different outcomes and choosing the most favorable option.
- **Machine Learning:** Probability theory plays a significant role in machine learning algorithms, particularly in the training and evaluation of models. For example, in supervised learning, probability distributions estimate the relationship between input features and output labels.
- **Natural Language Processing:** In natural language processing, probability theory models the likelihood of different word sequences, enabling AI systems to generate coherent sentences and understand the meaning behind human language.

In conclusion, understanding the fundamentals of probability theory is essential for anyone working in AI. Probability theory enables AI systems to make informed decisions, learn from data, and process human language by providing a mathematical framework to quantify uncertainty. As we continue to explore the vast landscape of AI, the importance of probability theory will only grow, making it an indispensable tool for developers and researchers alike.

Descriptive Statistics: Summarizing and Visualizing Data

In artificial intelligence, data is the lifeblood that fuels the development and growth of intelligent systems. AI practitioners must understand the underlying patterns and structures within the data to make informed decisions and build effective models. This is where descriptive statistics come into play, providing us with the tools to summarize and visualize data meaningfully.

Measures of Central Tendency

The first step in understanding data is to identify its central tendency, a single value representing the center or the "typical" value of a dataset. There are three primary measures of central tendency: the mean, the median, and the mode.

- The **mean**, often called the average, is calculated by adding up all the values in a dataset and dividing the sum by the total number of values. The mean is highly susceptible to the influence of outliers, which can skew the result.
- The **median** is the middle value in a dataset when the values are arranged in ascending or descending order. If there is an even number of values, the median is the average of the two middle values. The median is less affected by outliers and more accurately represents the central tendency in such cases.

- The **mode** is the value that occurs most frequently in a dataset. A dataset can have multiple modes or no modes at all. The mode is handy when dealing with categorical data, where the mean and median may not be applicable.

Measures of Dispersion

While central tendency provides a snapshot of the data's center, it is equally important to understand the spread or dispersion of the data. Measures of dispersion help us quantify the variability within a dataset.

- The **range** is the simplest measure of dispersion, calculated as the difference between a dataset's maximum and minimum values. While easy to compute, the range is highly sensitive to outliers.
- **Variance** is the average of the squared differences between each value and the mean. The standard deviation, the square root of the variance, is a more interpretable measure of dispersion expressed in the same units as the data. A larger standard deviation indicates greater variability in the data.
- The **Interquartile Range (IQR)** is the range within which the central 50% of the data lies. It is calculated as the difference between the first quartile (25th percentile) and the third quartile (75th percentile). The IQR is less sensitive to outliers and provides a more robust measure of dispersion.

Visualizing Data

Visual data representations can provide valuable insights and help identify patterns, trends, and outliers. Some standard data visualization techniques include:

- **Histograms:** A histogram is a graphical representation of the distribution of a dataset, where data is divided into bins or intervals, and the height of each bar represents the frequency of values within that bin.

- **Box Plots:** A box plot is a standardized way of displaying the distribution of a dataset based on the five-number summary (minimum, first quartile, median, third quartile, and maximum). It provides a visual representation of the data's central tendency, dispersion, and potential outliers.
- **Scatter Plots:** Scatter plots visualize the relationship between two continuous variables. Each point on the plot represents a single observation, with the x-axis representing one variable and the y-axis representing the other. Patterns in the scatter plot can indicate correlations or trends between the variables.

In conclusion, descriptive statistics and data visualization techniques are essential tools for understanding the structure and patterns within data. By effectively summarizing and visualizing data, AI practitioners can make informed decisions, identify potential issues, and build more accurate and robust models. As we delve deeper into AI, the importance of probability and statistics only grows, making these foundational concepts indispensable for success.

Inferential Statistics: Drawing Conclusions from Data

In artificial intelligence, making informed decisions based on data is crucial. This is where inferential statistics come into play. This section will delve into the fascinating world of inferential statistics, exploring how it enables us to draw conclusions from data and make predictions about future events or unknown parameters. We will also discuss the importance of hypothesis testing and confidence intervals in decision-making.

The Essence of Inferential Statistics

Inferential statistics is a branch of statistics that focuses on drawing conclusions about a population based on a sample of data. Unlike descriptive statistics, which merely summarize and visualize data, inferential statistics allow us to make predictions and generalizations about a larger group. This is particularly useful in AI, enabling machines to learn from limited data and make informed decisions.

Sampling and Sampling Distributions

The foundation of inferential statistics lies in the concept of sampling. A sample is a population subset selected to represent the entire group. On the other hand, sampling distributions describe the probability of obtaining different samples from the same population. Understanding sampling distributions is essential, as it allows us to quantify the uncertainty associated with our estimates and predictions.

Hypothesis Testing: The Art of Decision Making

One of the most critical aspects of inferential statistics is hypothesis testing. Hypothesis testing is a systematic method used to determine whether a claim about a population parameter is true or false. In AI, hypothesis testing can be employed to evaluate the performance of algorithms, compare different models, or validate assumptions.

The process of hypothesis testing involves the following steps:

- Formulate the null hypothesis (Ho) and the alternative hypothesis (H1).
- Choose a significance level (α), which represents the probability of rejecting the null hypothesis when it is true.
- Calculate the test statistic and the corresponding p-value.
- Compare the p-value with the significance level to make a decision.

Confidence Intervals: Quantifying Uncertainty

Another essential concept in inferential statistics is the confidence interval. A confidence interval is a range of values within which we expect the true population parameter to lie with a certain level of confidence. Confidence intervals measure the uncertainty associated with our estimates, allowing us to make more informed decisions in AI development.

The Power of Inferential Statistics in AI

In conclusion, inferential statistics play a vital role in developing and applying artificial intelligence. By enabling us to draw conclusions from data, make predictions, and quantify uncertainty, inferential statistics provide the foundation for informed decision-making in AI. As we continue to push the boundaries of AI technology, a solid understanding of probability and statistics will remain indispensable in creating intelligent systems that can learn from data and adapt to new situations.

Bayesian Inference: Updating Beliefs with Data

As we delve deeper into probability and statistics, it is crucial to understand the concept of Bayesian inference. This powerful tool allows us to update our beliefs based on new data. In artificial intelligence, Bayesian inference plays a significant role in decision-making, prediction, and modeling. In this section, we will explore the foundations of Bayesian inference, its applications in AI, and how it helps us make sense of uncertainty.

First, let us revisit the concept of conditional probability, which forms the basis of Bayesian inference. Conditional probability is the probability of an event occurring, given that another event has already occurred. In mathematical terms, the conditional probability of event A happening, given that event B has occurred, is denoted as $P(A|B)$. The vertical bar represents the phrase "given that."

Bayesian inference is built upon Bayes' theorem, a fundamental principle in probability theory that relates the conditional probabilities of two events. Mathematically, Bayes' theorem is expressed as:

$$P(A|B) = (P(B|A) * P(A)) / P(B)$$

In the context of AI, we can interpret A as a hypothesis or model and B as the observed data. The theorem allows us to update our belief in the hypothesis (P(A|B)) based on the likelihood of observing the data given the hypothesis (P(B|A)), the prior probability of the hypothesis (P(A)), and the probability of observing the data (P(B)).

Now, let's explore how Bayesian inference is applied in AI. One common application is in natural language processing (NLP), where Bayesian models are used to predict the next word in a sentence or classify a document's topic. For instance, a spam filter may use Bayesian inference to determine the probability that an email is spam, given the words it contains. By continuously updating the model with new data, the filter becomes more accurate in identifying spam emails.

Another application of Bayesian inference in AI is robotics, where robots use sensor data to update their beliefs about their environment. For example, a self-driving car may use Bayesian inference to estimate its position on the road, given the data from its sensors. As the car gathers more data, its position estimate becomes more accurate, allowing it to navigate safely and efficiently.

In conclusion, Bayesian inference is a powerful technique that enables AI systems to update their beliefs based on new data. By incorporating uncertainty and continuously refining their models, AI systems can make more informed decisions, predictions, and inferences. As we continue to develop and refine AI technologies, the role of probability and statistics, particularly Bayesian inference, will remain a critical component in our quest to create intelligent machines.

The Role of Probability and Statistics in AI Development

As we conclude this enlightening journey through probability and statistics, it is crucial to reflect on the significance of these mathematical concepts in the development of artificial intelligence. The intricate relationship between AI and the mathematical principles discussed in this chapter is undeniable, and understanding this connection is vital for anyone seeking to delve into the world of AI.

This chapter has explored the fundamentals of probability theory, descriptive and inferential statistics, and Bayesian inference. Each of these concepts plays a critical role in designing, implementing, and improving AI systems. By providing a solid foundation in these areas, we empower ourselves with the necessary tools to create intelligent systems that can learn from data, adapt to new information, and make informed decisions.

Probability theory is the backbone of AI, as it allows machines to reason under uncertainty. By quantifying the likelihood of various outcomes, AI systems can make informed decisions even when faced with incomplete or ambiguous information. This ability to reason probabilistically is essential for tasks such as natural language processing, computer vision, and robotics, where uncertainty is an inherent part of the problem.

Descriptive statistics, on the other hand, enable AI systems to summarize and visualize data effectively. By understanding the central tendency, dispersion, and distribution of data, AI developers can gain valuable insights into the underlying patterns and relationships within the data. This knowledge is crucial for designing algorithms that efficiently process and analyze large datasets, a common requirement in AI applications.

Inferential statistics play a pivotal role in AI by allowing developers to draw conclusions from data. Through hypothesis testing and confidence intervals, AI systems can make generalizations about populations based on samples, enabling them to learn from limited data and make predictions about future events. This ability to infer patterns and relationships from data is at the heart of machine learning, a subfield of AI that focuses on creating algorithms to learn from and make predictions based on data.

Lastly, Bayesian inference provides a powerful framework for updating beliefs with data. By combining prior knowledge with new information, AI systems can continuously refine their understanding of the world and make better decisions as they encounter new data. This dynamic learning process is particularly valuable in AI applications where the environment is constantly changing, such as autonomous vehicles or financial markets.

In conclusion, probability and statistics are indispensable tools in developing AI systems. By mastering these mathematical concepts, AI developers can create intelligent systems that can reason under uncertainty, learn from data, and adapt to new information. As AI advances and permeates various aspects of our lives, a strong foundation in probability and statistics will remain essential for those seeking to contribute to this exciting and rapidly evolving field.

Chapter Summary

- Probability and statistics are interconnected branches of mathematics that play a pivotal role in AI, providing a powerful toolkit for understanding and managing uncertainty.
- Probability theory focuses on the likelihood of events occurring and is essential for AI systems to make informed decisions based on data, particularly in natural language processing, computer vision, and robotics.
- Descriptive statistics help summarize and visualize data, enabling AI developers to gain valuable insights into the underlying patterns and relationships within the data, which is crucial for designing efficient algorithms.
- Inferential statistics allow AI systems to draw conclusions from data, predict future events, and generalize about populations based on samples, which is at the heart of machine learning.
- Bayesian inference is a powerful technique that enables AI systems to update their beliefs based on new data, allowing them to make more informed decisions, predictions, and inferences.

- Hypothesis testing and confidence intervals are essential components of inferential statistics, providing a systematic method for determining the validity of claims about population parameters and quantifying the uncertainty associated with estimates.
- AI applications such as natural language processing, robotics, and machine learning rely heavily on probability and statistics for decision-making, prediction, and modeling.
- A strong foundation in probability and statistics is essential for AI developers to create intelligent systems that can reason under uncertainty, learn from data, and adapt to new information.

3

CALCULUS: OPTIMIZING AI MODELS

W hen building artificial intelligence (AI), the ability to create efficient and accurate models is of utmost importance. The need for optimization becomes increasingly apparent as AI continues to permeate various industries and applications. One of the most powerful mathematical tools at our disposal for optimizing AI models is calculus. This branch of mathematics, which deals with the study of change and motion, has proven to be an invaluable asset in developing and fine-tuning AI systems.

At its core, calculus is the study of continuous change, and this concept makes it so well-suited for AI model optimization. AI models often involve complex mathematical functions that must be adjusted and fine-tuned to achieve the desired level of accuracy and efficiency. By applying the principles of calculus, we can gain a deeper understanding of these functions and how they can be manipulated to optimize the performance of our AI models.

A key concept in calculus that plays a crucial role in AI model optimization is the derivative. Derivatives allow us to analyze the rate of change of a function, which in turn helps us understand how minor adjustments to the input variables can impact the output of the model. This understanding is essential for gradient descent algorithms, widely used in AI to minimize model prediction errors.

As we delve deeper into calculus and AI model optimization, we will also explore the integration of calculus concepts in neural networks. Neural networks are a fundamental component of many AI systems, and understanding how calculus can be applied to these networks is essential for optimizing their performance.

Furthermore, we will examine the role of partial derivatives and multivariable optimization in AI models. As AI systems become more complex and involve a greater number of variables, the ability to optimize these models using multivariable calculus becomes increasingly important.

Throughout this chapter, we will also discuss real-world applications of calculus in AI models, showcasing the power and versatility of this mathematical tool in the context of AI optimization. From natural language processing to computer vision, calculus ensures that AI models are as efficient and accurate as possible.

In conclusion, the power of calculus in AI model optimization cannot be overstated. By understanding and applying the principles of calculus, we can unlock the full potential of AI systems and ensure that they continue to revolutionize the world around us. As we explore the various facets of calculus in AI model optimization, we will gain a deeper appreciation for the beauty and utility of this mathematical discipline in artificial intelligence.

The Role of Derivatives in Gradient Descent Algorithms

In artificial intelligence (AI), calculus is pivotal in optimizing AI models. One of the most significant applications of calculus in AI is using derivatives in gradient descent algorithms. In this section, we will delve into the importance of derivatives in these algorithms and explore how they contribute to optimizing AI models.

To begin with, let us first understand what gradient descent algorithms are. In simple terms, gradient descent is an optimization technique to minimize a function iteratively. It is widely employed in machine learning and deep learning for training AI models, particularly in scenarios that aim to minimize the error or loss function. The ultimate goal of gradient descent is to find the optimal set of parameters that minimize the loss function, thereby improving the model's performance.

Let us focus on derivatives and their role in gradient descent algorithms. In the context of calculus, a derivative represents the rate of change of a function concerning its input variable. In other words, it measures how sensitive the output of a function is to small changes in its input. In AI model optimization, derivatives are crucial in determining the direction and magnitude of the adjustments needed to minimize the loss function.

Gradient descent algorithms leverage the power of derivatives to update the model's parameters iteratively. The algorithm starts with an initial set of parameters and computes the gradient (i.e., the first-order derivative) of the loss function concerning each parameter. The gradient points in the direction of the steepest increase in the function's value. However, since our goal is to minimize the loss function, we move in the opposite direction of the gradient, taking steps proportional to the negative of the gradient.

A hyperparameter called the learning rate determines the size of the steps taken in the gradient descent algorithm. A smaller learning rate results in smaller steps, leading to a more gradual convergence to the optimal solution. Conversely, a larger learning rate may cause the algorithm to overshoot the optimal solution, potentially resulting in divergence. Striking the right balance in the learning rate is crucial for the success of the gradient descent algorithm.

In summary, derivatives play a vital role in gradient descent algorithms, widely used for optimizing AI models. By computing the gradient of the loss function, we can determine the direction and magnitude of the adjustments needed to minimize the loss function and improve the model's performance. The power of calculus, specifically the concept of derivatives, is indispensable in AI model optimization.

Integrating Calculus Concepts in Neural Networks

As we delve into the fascinating world of artificial intelligence, it becomes increasingly evident that calculus plays a pivotal role in optimizing AI models. One of the most prominent applications of calculus in AI is the integration of its concepts in neural networks. This section will explore how calculus concepts, particularly derivatives, and integrals, are employed in designing and optimizing neural networks.

Understanding Neural Networks

Before we discuss the role of calculus in neural networks, it is essential to understand what neural networks are and how they function. Neural networks are a series of interconnected nodes or neurons inspired by the human brain's structure. These networks are designed to recognize patterns, make decisions, and learn from experience. They consist of multiple layers, including an input layer, one or more hidden layers, and an output layer. Each neuron in a layer is connected to every neuron in the subsequent layer, with each connection having an associated weight.

The Role of Calculus in Neural Networks

In neural networks, the activation function is a crucial component that determines the output of a neuron based on its input. Common activation functions include the sigmoid, hyperbolic tangent (tanh), and rectified linear unit (ReLU) functions. The choice of activation function can significantly impact the network's performance and learning ability.

Calculus comes into play when we consider the derivatives of these activation functions. During the training process, neural networks use backpropagation to adjust the weights of the connections between neurons. This process involves computing the gradient of the error function with respect to each weight by using the chain rule, a fundamental concept in calculus. The chain rule requires the derivatives of the activation functions, making calculus an indispensable tool in optimizing neural networks.

Another essential aspect of neural networks is the loss function, which quantifies the difference between the network's predicted output and the actual target output. The goal of training a neural network is to minimize this loss function, and calculus plays a vital role in achieving this objective.

Gradient descent, a first-order optimization algorithm, is commonly used to minimize the loss function. It involves iteratively updating the weights of the connections in the neural network by moving in the direction of the steepest decrease of the loss function. This direction is determined by the gradient of the loss function, a vector of its partial derivatives with respect to each weight. Once again, calculus proves to be an essential tool in the optimization process.

In summary, calculus concepts are deeply ingrained in the design and optimization of neural networks. The use of derivatives in activation functions and the calculation of gradients in loss functions are just a few examples of how calculus plays a critical role in developing efficient and effective AI models. By understanding and applying these calculus concepts, AI practitioners can harness the power of neural networks to create innovative solutions to complex problems and drive advancements in artificial intelligence.

Partial Derivatives and Multivariable Optimization

In the realm of artificial intelligence, optimization is a crucial aspect of model development. As we delve deeper into calculus, we encounter partial derivatives and multivariable optimization, which significantly refine AI models. In this section, we will explore the concept of partial derivatives, their application in multivariable optimization, and how they contribute to the overall performance of AI models.

Understanding Partial Derivatives

To comprehend partial derivatives, we must first revisit the concept of a derivative. In single-variable calculus, a derivative represents the rate of change of a function concerning a single variable. However, in multivariable calculus, we deal with functions that have multiple input variables. This is where partial derivatives come into play.

Partial derivatives measure the rate of change of a multivariable function with respect to one variable while keeping the other variables constant. In other words, they allow us to examine how a function changes when we alter just one of its input variables. This is particularly useful in AI models, as they often involve multiple variables that interact with one another.

Multivariable Optimization Techniques

Multivariable optimization is finding the optimal values of multiple variables in a function to achieve a specific goal, such as minimizing error or maximizing efficiency. In AI models, this often involves adjusting the weights and biases of a neural network to improve its performance.

One popular technique for multivariable optimization is gradient descent, which we briefly touched upon in Section II. Gradient descent is an iterative algorithm that adjusts the variables of a function to minimize its output, such as the error in a neural network. It does this by computing the gradient, a vector of partial derivatives, and updating the variables accordingly.

Another technique is the Newton-Raphson method, which is an iterative method that uses second-order partial derivatives (also known as the Hessian matrix) to find the optimal values of a function. This method can be more efficient than gradient descent in certain situations, as it considers the curvature of the function, allowing for faster convergence to the optimal solution.

The Impact of Partial Derivatives and Multivariable Optimization on AI Models

Incorporating partial derivatives and multivariable optimization techniques into AI models can significantly improve their performance. By fine-tuning the variables within a model, we can minimize error and maximize efficiency, leading to more accurate predictions and better decision-making capabilities.

For instance, adjusting the weights and biases using gradient descent in a neural network can lead to a more accurate model that can better generalize to new data. Similarly, optimizing the parameters of an agent's policy in reinforcement learning can result in more effective decision-making and improved overall performance.

In conclusion, partial derivatives and multivariable optimization are essential tools in developing and refining AI models. By understanding and applying these concepts, we can create more efficient, accurate, and powerful artificial intelligence systems that profoundly impact various aspects of our lives.

Real-World Applications of Calculus in AI Models

In today's rapidly evolving technological landscape, artificial intelligence (AI) has become integral to various industries, from healthcare and finance to entertainment and transportation. The optimization of AI models is crucial for their performance, and calculus plays a pivotal role in achieving this. In this section, we will delve into some real-world applications of calculus in AI models, showcasing its significance in enhancing the efficiency and effectiveness of these systems.

Autonomous Vehicles

One of the most exciting applications of AI is in developing autonomous vehicles. These self-driving cars rely on many sensors and algorithms to navigate complex environments. Calculus comes into play in optimizing control algorithms, determining the vehicle's acceleration, braking, and steering. Engineers can fine-tune these algorithms using derivatives and integrals to ensure smooth and safe navigation, minimizing the risk of accidents and enhancing the overall driving experience.

Image Recognition and Computer Vision

AI-powered image recognition and computer vision systems have become increasingly prevalent in various applications, such as facial recognition, medical imaging, and surveillance. These systems rely on neural networks trained to identify patterns and features within images. Calculus is essential in optimizing these networks, as it helps adjust the weights and biases of the neurons during the training process. By employing gradient descent algorithms based on the concept of derivatives, these systems can learn to recognize and classify images with remarkable accuracy.

Natural Language Processing

Natural language processing (NLP) is another domain where AI has made significant strides, enabling machines to understand and generate human language. Applications like chatbots, voice assistants, and sentiment analysis rely on NLP algorithms to function effectively. Calculus is crucial in optimizing these algorithms, particularly in training recurrent neural networks (RNNs) and transformers, which are widely used in NLP tasks. By leveraging the power of partial derivatives and multivariable optimization, developers can fine-tune these models to comprehend better and respond to complex language patterns.

Financial Forecasting and Risk Management

AI has also found its way into finance, which is used to forecast market trends, manage investment portfolios, and assess risks. Calculus is instrumental in optimizing these AI models, as it enables the development of more accurate and reliable prediction algorithms. By using concepts such as derivatives and integrals, financial analysts can create models that can adapt to changing market conditions and make more informed decisions, ultimately leading to increased profitability and reduced risk.

In conclusion, the real-world applications of calculus in AI models are vast and varied, spanning numerous industries and domains. By harnessing the power of calculus, developers and engineers can optimize AI systems to perform at their peak, resulting in more accurate, efficient, and practical solutions to complex problems. As AI continues to advance and permeate our daily lives, the importance of calculus in optimizing these models will only grow, further solidifying its status as an essential mathematical tool in artificial intelligence.

The Power of Calculus in AI Model Optimization

As we reach the end of our exploration into the fascinating world of calculus and its applications in artificial intelligence, it is essential to take a moment to appreciate the power and significance of this mathematical discipline in optimizing AI models. Throughout this chapter, we have delved into the intricacies of calculus, from the role of derivatives in gradient descent algorithms to the integration of calculus concepts in neural networks. We have also examined the importance of partial derivatives and multivariable optimization, and real-world applications of calculus in AI models.

The power of calculus in AI model optimization cannot be overstated. It is the backbone of many optimization techniques that enable

AI models to learn and adapt to new data, ultimately improving their performance and accuracy. By understanding the underlying mathematical principles, we can better comprehend the inner workings of these models and develop more efficient and effective algorithms.

One of the most significant contributions of calculus to AI is its ability to help us understand and navigate the complex, multidimensional landscapes that AI models often inhabit. We can determine the direction and magnitude of change in these landscapes using derivatives and partial derivatives, allowing us to make informed decisions about adjusting our models to achieve optimal performance.

Furthermore, the integration of calculus concepts in neural networks has led to the development of powerful learning algorithms, such as backpropagation, which have revolutionized the field of AI and enabled the creation of sophisticated models capable of tackling a wide range of tasks. These models have found applications in various industries, from healthcare and finance to entertainment and transportation, transforming how we live, work, and interact with the world around us.

In conclusion, the power of calculus in AI model optimization is undeniable. It is a critical tool that allows us to harness the full potential of artificial intelligence, pushing the boundaries of what is possible and paving the way for a future where AI models continue to improve and evolve, ultimately leading to a more intelligent, efficient, and connected world. As we explore and develop new AI technologies, the importance of calculus and its applications in AI model optimization will continue to grow, solidifying its status as an indispensable component of the AI toolkit.

Chapter Summary

- Calculus is a powerful mathematical tool for optimizing AI models, as it deals with the study of continuous change, making it well-suited for understanding complex functions in AI systems.

- Derivatives play a crucial role in gradient descent algorithms, widely used in AI to minimize model prediction errors. They help determine the direction and magnitude of adjustments needed to optimize the model's performance.
- Calculus concepts, particularly derivatives and integrals, are integrated into neural networks, a fundamental component of many AI systems. Understanding how calculus can be applied to these networks is essential for optimizing their performance.
- Partial derivatives and multivariable optimization are important in AI models, as they allow for examining how a function changes when altering just one of its input variables. This is particularly useful in complex AI systems with multiple interacting variables.
- Gradient descent and the Newton-Raphson method are popular techniques for multivariable optimization, which rely on calculus concepts to find the optimal values of a function.
- Real-world applications of calculus in AI models span various industries, including autonomous vehicles, image recognition and computer vision, natural language processing, financial forecasting, and risk management.
- By harnessing the power of calculus, developers and engineers can optimize AI systems to perform at their peak, resulting in more accurate, efficient, and practical solutions to complex problems.
- As AI continues to advance and permeate our daily lives, the importance of calculus in optimizing AI models will only grow, further solidifying its status as an essential mathematical tool in artificial intelligence.

4

GRAPH THEORY: MODELING COMPLEX RELATIONSHIPS

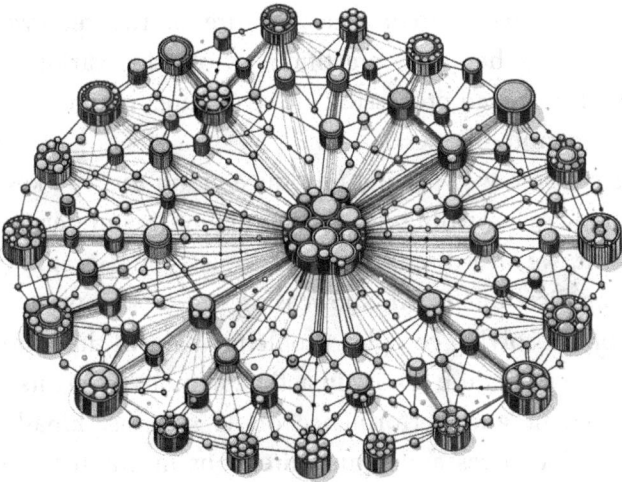

I n the ever-evolving landscape of artificial intelligence (AI), the ability to model and analyze complex relationships is crucial for developing intelligent systems that can effectively solve real-world problems. One of the most powerful mathematical tools for achieving this is graph theory, a branch of discrete mathematics that studies graphs. This chapter will delve into the fascinating world of graph theory and explore its significance in AI.

At its core, graph theory is the study of structures called graphs, which consist of vertices (also known as nodes) and edges (also known as links or connections) that connect these vertices. These graphs can represent various systems and relationships, from social networks and transportation systems to molecular structures and neural networks. By analyzing the properties and patterns within these graphs, we can gain valuable insights into the underlying systems they represent and develop more effective algorithms and AI models.

The importance of graph theory in AI cannot be overstated. As AI systems grow in complexity and sophistication, the need for efficient and accurate methods to model and analyze complex relationships becomes increasingly vital. Graph theory provides a versatile and robust framework for representing and processing information that closely mirrors the interconnected nature of the real world. This enables AI researchers and practitioners to tackle various problems, from natural language processing and computer vision to recommendation systems and robotics.

In this chapter, we will first introduce graph theory's key concepts and terminology, providing a solid foundation for understanding the subject. Next, we will discuss some of the essential graph algorithms and their applications in AI, showcasing the versatility and utility of graph theory in solving complex problems. We will then present real-world examples of graph theory in action, demonstrating its impact on the development and performance of AI systems. Finally, we will address the challenges and future directions in graph theory for AI, highlighting the potential for further advancements and breakthroughs in this exciting field.

As we embark on this journey through graph theory, we hope you will gain a deeper appreciation for its power and potential in advancing AI. By understanding and harnessing the capabilities of graph theory, we can unlock new possibilities for AI systems, paving the way for a future where intelligent machines can better understand and navigate the intricate web of relationships that define our world.

Key Concepts and Terminology in Graph Theory

As we delve into the fascinating world of graph theory, we must famil-
iarize ourselves with the key concepts and terminology that form the
foundation of this mathematical discipline. In this section, we will
explore the essential terms and ideas that will enable us better to
understand graph theory's applications in artificial intelligence.

A graph is a mathematical representation of a set of objects, called
vertices (or nodes), and the relationships between them, called edges
(or links). In AI, vertices can represent entities such as people, places, or
things, while edges represent the relationships or connections between
these entities.

Graphs can be classified into two main types based on the nature of
their edges. In an undirected graph, edges have no specific direction,
implying that the relationship between vertices is bidirectional.
Conversely, a directed graph (or digraph) has edges with a specific
direction, indicating that the relationship between vertices is unidi-
rectional.

Another classification of graphs is based on the presence or
absence of weights assigned to their edges. In a weighted graph, each
edge is assigned a numerical value, representing the strength or cost of
the connection between the vertices. An unweighted graph, on the
other hand, does not have any values assigned to its edges.

The degree of a vertex is the number of edges connected to it. In a
directed graph, we distinguish between the in-degree (number of
incoming edges) and the out-degree (number of outgoing edges) of a
vertex.

A path in a graph is a sequence of vertices connected by edges. The
length of a path is the number of edges in the sequence. In a weighted
graph, the weight of a path is the sum of the weights of its edges.

A cycle is a path that starts and ends at the same vertex, with no
other vertices repeated in the sequence. A graph is called acyclic if it
contains no cycles.

A graph is connected if there is a path between every pair of
vertices. In a directed graph, we distinguish between strongly

connected (a path exists in both directions between every pair of vertices) and weakly connected (the graph is connected if we ignore the direction of the edges) graphs.

A subgraph is a graph formed by a subset of the vertices and edges of a larger graph. A subgraph is induced if it contains all the edges of the larger graph that connect the selected vertices.

A tree is a connected, acyclic graph. A forest is a collection of trees, i.e., a graph that is acyclic but not necessarily connected.

Two graphs are isomorphic if a one-to-one correspondence exists between their vertices and edges, such that the connectivity between vertices is preserved.

With these fundamental concepts and terminology in mind, we are better equipped to explore the various graph algorithms and their applications in artificial intelligence. As we progress through this chapter, we will see how graph theory plays a pivotal role in modeling complex relationships and solving intricate problems in AI.

Graph Algorithms and Their Applications in AI

In artificial intelligence (AI), graph theory plays a pivotal role in modeling complex relationships and solving intricate problems. Graph algorithms, the mathematical procedures designed to analyze and manipulate graph structures, are indispensable tools for AI researchers and practitioners. In this section, we will delve into the world of graph algorithms, exploring their applications in AI and how they contribute to the development of intelligent systems.

Graph algorithms allow us to traverse, search, and analyze graphs systematically and efficiently. These algorithms can be broadly classified into two categories: traversal algorithms and optimization algorithms. Traversal algorithms, such as depth-first search (DFS) and breadth-first search (BFS), are used to explore the vertices and edges of a graph, while optimization algorithms, like Dijkstra's and Bellman-Ford, are employed to find the shortest path between two nodes or solve other optimization problems.

Applications of Graph Algorithms in AI

Natural Language Processing (NLP): Graph algorithms are widely used in NLP to model and analyze the relationships between words, phrases, and sentences. For instance, the PageRank algorithm, initially developed by Google to rank web pages, can be applied to text corpora to identify the most important words or phrases in a document. This technique is beneficial in keyword extraction, summarization, and sentiment analysis tasks.

Computer Vision: In computer vision, graph algorithms are employed to model the relationships between pixels, regions, and objects in an image. For example, the minimum spanning tree (MST) algorithm can segment an image into distinct regions based on color or texture similarity. Additionally, graph-based techniques like spectral clustering can be applied to group similar images together, aiding in object recognition and scene understanding tasks.

Social Network Analysis: AI systems that analyze social networks rely heavily on graph algorithms to uncover patterns and trends in individual relationships. Algorithms like community detection and betweenness centrality can help identify influential users, detect tightly-knit groups, and predict the spread of information or influence through a network.

Recommender Systems: Graph algorithms play a crucial role in developing recommender systems, which provide personalized suggestions to users based on their preferences and behavior. Collaborative filtering, a popular technique in recommender systems, utilizes graph algorithms to identify users with similar tastes and recommend items based on their collective preferences.

The Impact of Graph Algorithms on AI: The applications of graph algorithms in AI are vast and varied, enabling researchers and practitioners to model complex relationships and solve intricate problems easily. By leveraging the power of graph algorithms, AI systems can better understand and interpret the world around them, leading to more accurate predictions, improved decision-making, and enhanced user experiences.

In conclusion, graph algorithms are indispensable tools in the AI toolbox, providing valuable insights and solutions to various challenges. As AI advances and evolves, the importance of graph theory and its associated algorithms will only grow, paving the way for more sophisticated and intelligent systems.

Challenges and Future Directions in Graph Theory for AI

As we delve deeper into the world of artificial intelligence, the significance of graph theory in modeling complex relationships becomes increasingly apparent. However, despite its numerous applications and potential, challenges still need to be addressed, and future directions need to be explored to fully harness the power of graph theory in advancing AI.

One of the primary challenges in applying graph theory to AI is the sheer size and complexity of real-world graphs. As the number of nodes and edges in a graph increases, so does the computational complexity of graph algorithms. This can lead to scalability issues, mainly when dealing with massive datasets common in AI applications. Researchers continuously develop more efficient algorithms and parallel processing techniques to overcome these scalability challenges.

Another challenge lies in the dynamic nature of many real-world graphs. In many AI applications, the relationships between entities are not static but change over time. This requires the development of dynamic graph algorithms that can adapt to these changes and still provide accurate and timely insights. Additionally, incorporating uncertainty and probabilistic reasoning into graph models is an area that needs further exploration, as real-world relationships often involve a degree of uncertainty.

In terms of future directions, one promising area of research is integrating graph theory with other AI techniques, such as machine learning and natural language processing. By combining the strengths of these different approaches, we can develop more powerful and versatile AI systems that tackle a wider range of problems. For example, graph-based machine learning algorithms can be used to analyze and learn from the structure of large-scale networks, while natural language processing can help extract meaningful relationships from unstructured text data.

Another exciting direction is the development of new graph-based AI models that can better capture the complexity of real-world relationships. For instance, hypergraphs, which allow edges to connect more than two nodes, can be used to model higher-order relationships and interactions between entities. Similarly, multilayer graphs can be employed to represent multiple types of relationships or different aspects of a single relationship, providing a richer and more nuanced understanding of complex systems.

In conclusion, graph theory holds immense potential in advancing the field of artificial intelligence. By addressing the challenges of scalability, dynamic relationships, and uncertainty and exploring new directions, such as integrating graph theory with other AI techniques and developing more sophisticated graph models, we can unlock the full power of graph theory in AI.

The Power of Graph Theory in Advancing AI

As we have explored throughout this chapter, graph theory plays a pivotal role in developing and advancing artificial intelligence. By providing a robust mathematical framework for modeling complex relationships, graph theory enables AI systems to understand better, analyze, and navigate the intricate web of connections within various domains. In this conclusion, we will briefly recap the key points discussed in this chapter and highlight the immense potential of graph theory in propelling AI to new heights.

We began our journey into graph theory by introducing its fundamental concepts and terminology, such as vertices, edges, and adjacency matrices. These building blocks serve as the foundation for more advanced topics and applications in AI. We then delved into graph algorithms, essential tools for AI systems to process and analyze graph data efficiently. These algorithms, such as Dijkstra's shortest path and the PageRank algorithm, have been instrumental in solving complex problems in AI, from natural language processing to social network analysis.

Throughout the chapter, we showcased real-world examples of graph theory in action within AI systems. These examples illustrated the versatility and power of graph theory in addressing diverse challenges, such as detecting fraud in financial transactions, predicting protein interactions in bioinformatics, and optimizing transportation networks. These applications underscore the immense value of graph theory in enhancing the capabilities of AI systems across various industries and disciplines.

However, we also acknowledged the challenges and future directions in graph theory for AI. As AI systems evolve and tackle increasingly complex problems, researchers must develop novel graph algorithms and techniques to keep pace with these advancements. Moreover, the ethical implications of AI systems that rely on graph theory, such as privacy concerns and potential biases in graph data, must be carefully addressed.

In conclusion, the power of graph theory in advancing AI cannot be overstated. Graph theory has become an indispensable tool in the AI toolkit by enabling AI systems to model and navigate complex relationships. As we continue to push the boundaries of what AI can achieve, graph theory will undoubtedly remain a critical component in the ongoing quest to develop intelligent systems that can solve the most pressing challenges of our time.

Chapter Summary

- Graph theory is a branch of discrete mathematics that deals with the study of graphs, which consist of vertices (nodes) and edges (connections). It is crucial for modeling and analyzing complex relationships in artificial intelligence (AI).

- Key concepts and terminology in graph theory include graphs, directed and undirected graphs, weighted and unweighted graphs, degree, path, cycle, connectivity, subgraph, trees, forests, and graph isomorphism.
- Graph algorithms are essential tools for AI systems to process and analyze graph data efficiently. They can be broadly classified into traversal algorithms (e.g., depth-first search and breadth-first search) and optimization algorithms (e.g., Dijkstra's and Bellman-Ford).
- Graph theory has numerous applications in AI, including natural language processing (NLP), computer vision, social network analysis, recommender systems, and autonomous vehicles and robotics.
- Real-world examples of graph theory in AI systems include social network analysis, recommendation systems, natural language processing, computer vision, autonomous vehicles, and robotics.
- Challenges in applying graph theory to AI include the size and complexity of real-world graphs, the dynamic nature of many real-world graphs, and incorporating uncertainty and probabilistic reasoning into graph models.
- Future directions in graph theory for AI include the integration of graph theory with other AI techniques (e.g., machine learning and NLP), the development of new graph-based AI models (e.g., hypergraphs and multilayer graphs), and addressing ethical implications of AI systems that rely on graph theory.
- Graph theory holds immense potential in advancing the field of AI by providing a robust mathematical framework for modeling complex relationships, enabling AI systems to understand better, analyze, and navigate the intricate web of connections within various domains.

5

DISCRETE MATHEMATICS: EXPLORING COMBINATORIAL PROBLEMS

I n the world of artificial intelligence (AI), the need for a solid mathematical foundation cannot be overstated. Discrete mathematics is one of the most crucial branches of mathematics that plays a pivotal role in AI development. This chapter aims to provide an insightful overview of discrete mathematics, its core concepts, and its significance in AI.

Discrete mathematics, as the name suggests, deals with discrete or distinct objects, as opposed to continuous mathematics, which focuses on continuous objects. In simpler terms, discrete mathematics is the study of countable, separate values, while continuous mathematics deals with values that can vary continuously. The former is particularly relevant to AI, as it helps understand and solve problems involving finite or countable structures, such as networks, algorithms, and data manipulation.

The importance of discrete mathematics in AI stems from its ability to model and analyze complex relationships, optimize problem-solving, and make informed decisions. As AI systems are designed to process and interpret vast amounts of data, discrete mathematics provides the necessary tools to handle these tasks efficiently and accurately. Moreover, discrete mathematics is the backbone of many AI algorithms, as it enables the development of efficient and reliable solutions to a wide range of problems.

One of the critical aspects of discrete mathematics that makes it indispensable for AI is its focus on combinatorial problems. Combinatorial problems involve finding, counting, or optimizing arrangements of objects under specific constraints. These problems are ubiquitous in AI, arising in various applications, such as natural language processing, computer vision, and machine learning.

This chapter will delve into the fascinating world of discrete mathematics and explore its various concepts, such as permutations, combinations, and partitions. We will also discuss graph theory, which is instrumental in modeling and analyzing complex relationships in AI systems. Furthermore, we will examine probability theory, which plays a crucial role in making informed decisions in AI, and algorithmic efficiency, which is vital for optimizing combinatorial problem-solving.

By the end of this chapter, you will have a comprehensive understanding of the power of discrete mathematics in AI development and its potential to revolutionize how we approach and solve complex problems. So, let us embark on this exciting journey and unravel the mysteries of discrete mathematics and its applications in AI.

Combinatorial Concepts: Permutations, Combinations, and Partitions

In artificial intelligence, discrete mathematics is pivotal in solving complex problems and making informed decisions. One of the most fascinating aspects of discrete mathematics is combinatorial concepts, which involve the study of finite or countable discrete structures. This section will delve into the world of permutations, combinations, and partitions, exploring their significance in AI development.

Permutations: Ordering Matters

Permutations are a fundamental concept in combinatorial mathematics, which deals with the arrangement of objects in a specific order. In AI, permutations are often used to analyze and solve problems that involve sequencing, such as determining the optimal route for a delivery truck or finding the best sequence of moves in a game.

A permutation is an arrangement of objects in a specific order without repetition. The number of permutations of a set of n objects is given by the factorial function, denoted as n! (n factorial). For example, the number of permutations of a set of 3 objects (A, B, C) is $3! = 3 \times 2 \times 1 = 6$. These permutations are ABC, ACB, BAC, BCA, CAB, and CBA.

Combinations: Order Doesn't Matter

While permutations focus on the arrangement of objects in a specific order, combinations are concerned with the selection of objects without regard to their arrangement. Combinations are used in AI to solve problems that involve selecting a subset of items from a larger set, such as choosing the best features for a machine learning model or selecting a group of sensors for optimal data collection.

The number of combinations of a set of n objects taken r at a time, denoted as C(n, r) or "n choose r," is calculated using the formula:

$$C(n, r) = n! / (r! \, (n-r)!)$$

For example, the number of combinations of a set of 3 objects (A, B, C) taken 2 at a time is C(3, 2) = 3! / (2! (3-2)!) = 3. These combinations are AB, AC, and BC.

Partitions: Dividing into Distinct Groups

Partitions, another essential combinatorial concept, involve dividing a set of objects into distinct, non-overlapping groups. Partitions are used in AI to solve problems that require dividing data into clusters or groups, such as image segmentation or customer segmentation in marketing.

The number of partitions of a set of n objects into k non-empty groups is denoted as P(n, k) and can be calculated using the Stirling numbers of the second kind or recursive formulas. For example, the number of partitions of a set of 4 objects (A, B, C, D) into 2 non-empty groups is P(4, 2) = 7. These partitions are (AB, CD), (AC, BD), (AD, BC), (ABC, D), (ABD, C), (ACD, B), and (BCD, A).

In conclusion, combinatorial concepts such as permutations, combinations, and partitions are indispensable tools in developing AI systems. By understanding and applying these concepts, AI developers can efficiently solve complex problems, optimize decision-making processes, and ultimately create more intelligent and effective AI solutions.

Graph Theory: Modeling and Analyzing Complex Relationships

In artificial intelligence, the ability to model and analyze complex relationships is crucial for developing intelligent systems that can navigate and make sense of the intricate web of connections in the real world. Graph theory, a branch of discrete mathematics, provides a robust framework for representing and understanding these relationships. In this section, we will delve into the fundamentals of graph theory, explore its applications in AI, and discuss how it can be used to solve combinatorial problems.

At its core, graph theory studies mathematical structures called graphs, consisting of vertices (nodes) and edges (links) connecting these vertices. Graphs can be used to represent a wide variety of relationships, such as social networks, transportation systems, and even the structure of the Internet. There are two main types of graphs: directed and undirected. In directed graphs, the edges have a specific direction, indicating a one-way relationship between vertices. In contrast, undirected graphs have edges without direction, signifying a mutual relationship between vertices.

To work with graphs, it is essential to understand their representation and the terminology associated with them. Graphs can be represented using adjacency matrices, adjacency lists, or edge lists. An adjacency matrix is a square matrix where the entry at position (i, j) indicates the presence of an edge between vertices i and j. Adjacency lists store the neighbors of each vertex in a list, while edge lists store all the edges in a graph as pairs of vertices.

Some common terms in graph theory include:

- **Degree:** The number of edges connected to a vertex.
- **Path:** A sequence of vertices connected by edges.
- **Cycle:** A path that starts and ends at the same vertex without repeating other vertices.
- **Connected graph:** A graph with a path between every pair of vertices.
- **Subgraph:** A graph formed by a subset of vertices and edges from a larger graph.
- **Tree:** A connected graph without cycles.

Graph theory has numerous applications in AI, ranging from natural language processing to computer vision. Some examples include:

- **Social network analysis:** Graphs can be used to model relationships between individuals in a social network, enabling AI systems to identify influential users, detect communities, and predict the spread of information.

- **Recommendation systems:** By representing users and items as vertices and their interactions as edges, AI algorithms can analyze the graph's structure to make personalized recommendations.
- **Image segmentation:** In computer vision, graphs can model the relationships between pixels in an image, allowing AI systems to identify and separate distinct objects.
- **Pathfinding:** Graphs can represent the layout of physical spaces, such as road networks or game maps, enabling AI agents to find the shortest or most efficient path between two points.

Graph theory provides a rich toolbox for tackling combinatorial problems in AI. For example, the traveling salesperson problem, which involves finding the shortest route that visits a set of cities and returns to the starting city, can be modeled as a graph with cities as vertices and distances as edge weights. Similarly, the maximum flow problem, which seeks to maximize the flow of resources through a network, can be represented as a directed graph with capacities assigned to each edge.

By leveraging graph algorithms and optimization techniques, AI developers can efficiently solve these and other combinatorial problems, unlocking the full potential of discrete mathematics in AI development.

In conclusion, graph theory plays a vital role in modeling and analyzing complex relationships in artificial intelligence. By understanding the fundamentals of graph theory and its applications in AI, developers can harness its power to create intelligent systems capable of navigating and making sense of the intricate connections that define our world.

Probability Theory: Making Informed Decisions in AI Systems

As we delve deeper into discrete mathematics, we encounter probability theory, a branch that plays a crucial role in developing and functioning artificial intelligence (AI) systems. Probability theory allows us to make informed decisions by quantifying the likelihood of various outcomes. This section will explore the fundamentals of probability theory and its applications in AI systems.

Probability theory is a mathematical framework that deals with the analysis of random phenomena. It provides us with the tools to model and predict the behavior of complex systems under uncertainty. In AI, probability theory represents and manipulates uncertain information, enabling machines to reason and make decisions in the face of incomplete or ambiguous data.

One of the key concepts in probability theory is the notion of a random variable. A random variable is a function that assigns a numerical value to each outcome of a random experiment. For example, in a coin toss experiment, we can define a random variable X that takes the value 1 if the coin lands heads and 0 if it lands tails. The probability distribution of a random variable describes the likelihood of each possible outcome. In our coin toss example, if the coin is fair, the probability distribution of X would assign a probability of 0.5 to both 1 (heads) and 0 (tails).

In AI systems, we often deal with multiple random variables related to each other. For instance, consider a robot navigating through an environment with obstacles. The robot's position, the positions of the obstacles, and the robot's sensor readings can all be modeled as random variables. We use joint probability distributions to reason about the relationships between these variables, which describe the likelihood of different combinations of variable values.

Conditional probability is another essential concept in probability theory. It allows us to update our beliefs about a random variable based on new information. In the context of AI, this is particularly useful for incorporating sensor readings or user input into the system's decision-making process. Bayes' theorem, a fundamental result in probability theory, provides a way to compute conditional probabilities by relating them to the joint probabilities of the variables involved.

AI systems often need to make decisions based on uncertain information. To do this, they employ decision theory, combining probability and utility theories to determine the best course of action. In decision theory, an AI system evaluates the expected utility of each possible action, considering both the likelihood of different outcomes and the desirability of those outcomes. The system then selects the action with the highest expected utility.

In conclusion, probability theory is a powerful tool for modeling and reasoning about uncertainty in AI systems. By representing and manipulating uncertain information, AI systems can make informed decisions even in the face of incomplete or ambiguous data. As we continue developing increasingly sophisticated AI technologies, our understanding and application of probability theory will remain vital to their success.

Algorithmic Efficiency: Optimizing Combinatorial Problem Solving

In the realm of artificial intelligence, the ability to solve complex combinatorial problems efficiently is of paramount importance. Optimizing algorithms becomes even more critical as AI systems evolve and tackle increasingly intricate tasks. In this section, we will delve into the concept of algorithmic efficiency, explore its significance in combinatorial problem-solving, and discuss various techniques to enhance the performance of AI systems.

Understanding Algorithmic Efficiency

Algorithmic efficiency refers to the effectiveness of an algorithm in terms of the resources it consumes, such as time and memory, to solve a given problem. In the context of combinatorial problems, efficient algorithms can quickly find optimal solutions while minimizing the computational resources required. The efficiency of an algorithm is often measured using its time complexity, which is a function of the input size that describes the number of basic operations the algorithm performs.

The Importance of Algorithmic Efficiency in AI

As AI systems tackle more complex problems, efficient algorithms become increasingly crucial. Inefficient algorithms can lead to slow processing times, excessive memory usage, and the inability to find optimal solutions within a reasonable timeframe. This can be particularly detrimental in real-world applications, where timely decision-making is often essential. By optimizing algorithmic efficiency, AI developers can create systems that are more responsive, accurate, and capable of handling large-scale combinatorial problems.

Techniques for Optimizing Combinatorial Problem Solving

There are several techniques that can be employed to optimize the efficiency of combinatorial problem-solving algorithms. Some of these include:

- **Divide and Conquer:** This approach involves breaking a problem down into smaller subproblems, solving each subproblem independently, and then combining the solutions to form the overall solution. By tackling smaller, more manageable tasks, the algorithm can arrive at a solution more quickly.
- **Dynamic Programming:** This technique involves solving a problem by breaking it down into overlapping subproblems and storing the results of these subproblems to avoid redundant computations. By reusing previously computed results, dynamic programming can significantly reduce the time complexity of an algorithm.
- **Greedy Algorithms:** Greedy algorithms make locally optimal choices at each step to find a globally optimal solution. While they may not always yield the best possible solution, they can often provide a good approximation relatively quickly.

- **Heuristics:** Heuristic methods involve using problem-specific knowledge or intuition to guide the search for a solution. By incorporating domain-specific insights, heuristics can focus the search on more promising areas of the solution space, thereby reducing the time required to find an optimal solution.

The Power of Discrete Mathematics in AI Development

In conclusion, algorithmic efficiency is vital in developing AI systems capable of solving complex combinatorial problems. By employing techniques such as divide and conquer, dynamic programming, greedy algorithms, and heuristics, AI developers can create systems that are more responsive, accurate, and capable of handling large-scale problems. As the field of AI continues to advance, the importance of discrete mathematics, particularly the optimization of combinatorial problem-solving algorithms, will only continue to grow.

In this final section, we shall reflect on the significance of discrete mathematics in artificial intelligence (AI) and how it empowers developers to create innovative and efficient solutions. Throughout this chapter, we have delved into the various aspects of discrete mathematics, including combinatorial concepts, graph theory, probability theory, and algorithmic efficiency. Each of these components plays a crucial role in developing and optimizing AI systems, enabling them to tackle complex problems and make informed decisions.

Discrete mathematics is the foundation for understanding and analyzing AI systems' intricate structures and relationships. By employing combinatorial concepts such as permutations, combinations, and partitions, developers can explore the vast possibilities and configurations that arise in AI problems. This, in turn, allows them to design algorithms that can efficiently navigate through these possibilities and arrive at optimal solutions.

Graph theory, another essential aspect of discrete mathematics, enables developers to model and analyze complex relationships within AI systems. By representing these relationships as graphs, developers can gain valuable insights into the structure and behavior of their AI systems. This understanding is vital for designing algorithms that can effectively traverse and manipulate these relationships, leading to more robust and adaptable AI solutions.

Probability theory plays a significant role in AI development by providing a framework for making informed decisions in uncertain environments. AI systems often need to make choices based on incomplete or noisy data, and probability theory equips them with the tools to do so in a principled manner. By incorporating probabilistic reasoning into their algorithms, developers can create AI systems capable of making intelligent decisions even when uncertain.

Algorithmic efficiency is a critical consideration in AI development, as it directly impacts the performance and scalability of AI systems. Discrete mathematics offers a wealth of techniques for optimizing combinatorial problem-solving, allowing developers to create algorithms that can efficiently tackle complex problems. By harnessing the power of discrete mathematics, developers can ensure that their AI systems are effective and computationally efficient.

In conclusion, discrete mathematics is an indispensable tool in developing AI systems. Its various components, including combinatorial concepts, graph theory, probability theory, and algorithmic efficiency, provide developers with the necessary knowledge and techniques to create innovative and powerful AI solutions. As AI continues to advance and permeate various aspects of our lives, the importance of discrete mathematics in AI development will only grow, making it an essential area of study for aspiring AI developers and researchers.

Chapter Summary

- Discrete mathematics is a crucial branch of mathematics in AI development, as it helps understand and solve problems involving finite or countable structures, such as networks, algorithms, and data manipulation.

- Combinatorial concepts, including permutations, combinations, and partitions, are essential tools in AI development, enabling efficient problem-solving, optimization, and decision-making processes.
- Graph theory plays a vital role in modeling and analyzing complex relationships in AI systems, allowing developers to create intelligent systems capable of navigating and making sense of intricate connections.
- Probability theory is a powerful tool for modeling and reasoning about uncertainty in AI systems, enabling machines to make informed decisions even in the face of incomplete or ambiguous data.
- Algorithmic efficiency is paramount in AI development, as it directly impacts the performance and scalability of AI systems. Optimizing algorithms ensures that AI systems are effective and computationally efficient.
- Techniques for optimizing combinatorial problem-solving include divide and conquer, dynamic programming, greedy algorithms, and heuristics. These methods can help AI developers create more responsive, accurate, and capable systems.
- Decision theory, which combines probability theory with utility theory, is used in AI systems to evaluate the expected utility of each possible action and select the action with the highest expected utility.
- As AI continues to advance and permeate various aspects of our lives, the importance of discrete mathematics in AI development will only grow, making it an essential area of study for aspiring AI developers and researchers.

6

NUMERICAL METHODS: SOLVING EQUATIONS AND APPROXIMATING FUNCTIONS

In the constantly-changing world of artificial intelligence (AI), the ability to solve complex problems and make accurate predictions is paramount. As AI systems advance, they increasingly rely on mathematical models and algorithms to process vast amounts of data and make informed decisions. One of the key components in developing these AI systems is using numerical methods, which are techniques for solving mathematical problems using numerical approximations.

Numerical methods play a crucial role in AI, as they provide the foundation for many algorithms and models used in machine learning, data analysis, and optimization. These methods enable AI systems to tackle problems that are too complex or time-consuming to solve analytically, allowing for more efficient and accurate solutions.

This chapter will delve into the fascinating world of numerical methods and their applications in AI. We will explore various techniques for solving equations, approximating functions and optimizing solutions, all essential tools in the AI developer's toolkit.

First, we will examine root-finding techniques used to solve equations and find the points where a given function equals zero. These methods are fundamental in AI, as they can be applied to a wide range of problems, from optimizing neural networks to solving systems of linear equations.

Next, we will discuss interpolation and curve fitting, which are techniques for approximating functions based on a set of data points. These methods are instrumental in AI for tasks such as data analysis, pattern recognition, and function approximation, as they allow for creating smooth and continuous functions that can be easily analyzed and manipulated.

Following this, we will explore numerical integration and differentiation, which are essential tools for analyzing and understanding the behavior of functions. These techniques are invaluable in AI, as they enable the calculation of areas, volumes, and rates of change, all of which are critical for understanding the dynamics of complex systems.

Finally, we will delve into optimization techniques used to find the best possible solution to a given problem. In AI, optimization is a critical component of many algorithms, as it allows for fine-tuning models and discovering optimal solutions to complex problems.

In conclusion, numerical methods are an indispensable part of AI development, providing the mathematical foundation for many algorithms and models that drive modern AI systems. By understanding and mastering these techniques, AI developers can unlock the full potential of their creations and push the boundaries of what is possible in the realm of artificial intelligence.

Root Finding Techniques for Solving Equations

In the fascinating artificial intelligence (AI) world, numerical methods play a crucial role in solving complex mathematical problems that arise in various applications. One such essential aspect is finding the roots of equations, the foundation for many AI algorithms. In this section, we will delve into the realm of root-finding techniques, exploring their significance and how they contribute to the development of AI systems.

Understanding Roots of Equations

Before diving into the techniques, let's first understand the roots of equations. In mathematics, a root of an equation is a value that makes it true when substituted into the equation. In other words, it is the point where the function intersects the x-axis. Finding the roots of an equation is a fundamental problem in mathematics and has numerous applications in AI, such as optimization, computer graphics, and machine learning.

Bracketing Methods

Bracketing methods are a class of root-finding techniques that involve isolating the root within an interval and successively narrowing it down. These methods guarantee convergence, provided the function is continuous, and the initial interval contains a root. Two popular bracketing methods are:

- **Bisection Method:** This technique involves dividing the interval into two equal parts and determining which half contains the root. The process is repeated until the desired level of accuracy is achieved. The bisection method is simple, reliable, and easy to implement, but it can be relatively slow compared to other methods.

- **Regula Falsi (False Position) Method:** This method is an improvement over the bisection method. Instead of dividing the interval into two equal parts, it uses a linear interpolation between the function values at the interval endpoints to estimate the root. This approach generally converges faster than the bisection method but may sometimes be slower due to specific function properties.

Iterative Methods

Iterative methods are another class of root-finding techniques that involve refining an initial guess of the root through a series of iterations. These methods can be faster than bracketing methods but may only sometimes guarantee convergence. Some widely used iterative methods are:

- **Newton-Raphson Method:** This technique uses the function's derivative to approximate the root. It involves linearizing the function at the initial guess and finding the intersection point with the x-axis. The process is repeated until convergence is achieved. The Newton-Raphson method is known for its fast convergence rate but may fail if the initial guess is not close enough to the root or if the function's derivative is zero.

- **Secant Method:** The secant method is a modification of the Newton-Raphson method that does not require the derivative of the function. Instead, it uses two initial guesses and approximates the derivative using the finite difference between the function values. The secant method converges faster than the bisection method but may be slower than the Newton-Raphson method.

In conclusion, root-finding techniques are indispensable tools in the AI developer's arsenal, enabling them to solve complex equations that arise in various applications. By understanding and implementing these methods, AI systems can be developed more efficiently and accurately, paving the way for groundbreaking innovations.

Interpolation and Curve Fitting for Function Approximation

In artificial intelligence (AI), the ability to approximate functions is a crucial skill that enables machines to learn from data and make predictions. Interpolation and curve fitting are two powerful techniques that allow us to create mathematical models that closely mimic real-world phenomena' behavior. This section will delve into the fascinating world of function approximation, exploring the concepts of interpolation and curve fitting and discussing their significance in AI development.

Understanding Interpolation

Interpolation is a mathematical technique used to estimate the value of a function at a specific point, given a set of known data points. The primary goal of interpolation is to construct a smooth curve that passes through all the given data points, thereby providing a means to predict the function's behavior in between these points. This is particularly useful in AI when we have limited data and need to make inferences about the underlying function.

There are several interpolation methods, each with its strengths and weaknesses. Some of the most common techniques include:

- **Linear Interpolation:** This method connects adjacent data points with straight lines, creating a piecewise linear function. It is simple to implement and computationally efficient but may only accurately capture the true behavior of the function if it is linear between the given points.
- **Polynomial Interpolation:** This technique involves fitting a polynomial function to the data points. The degree of the polynomial is typically chosen to be one less than the

number of data points, ensuring the curve passes through all of them. While this method can provide a smooth and continuous curve, it may suffer from oscillations and overfitting, especially when dealing with a large number of data points.

- **Spline Interpolation:** This method divides the data into intervals and fits a separate polynomial function to each interval. The polynomials are chosen to be smooth and continuous at the interval boundaries. Spline interpolation balances simplicity and accuracy, making it a popular choice for many applications.

Curve Fitting: A Generalization of Interpolation

While interpolation focuses on constructing a curve that passes through all the given data points, curve fitting takes a more generalized approach. Curve fitting aims to find a function that best approximates the underlying relationship between the variables, even if it does not pass through every data point. This is particularly useful in AI when dealing with noisy or imperfect data, as it allows us to capture the overall trend without being overly influenced by individual outliers.

Curve fitting typically involves selecting a suitable function (e.g., linear, polynomial, exponential) and adjusting its parameters to minimize the discrepancy between the predicted and actual values. This process is often achieved using optimization techniques, such as least squares or gradient descent.

The Role of Interpolation and Curve Fitting in AI

Interpolation and curve fitting play a vital role in developing AI systems. They enable machines to learn from data, generalize patterns, and predict unseen instances. Some of the key applications of these techniques in AI include:

- **Regression Analysis:** In supervised learning, interpolation and curve fitting can be used to create regression models that predict the output of a continuous variable based on one or more input features.
- **Image Processing:** These techniques can be employed to enhance image resolution, fill in missing pixels, or correct distortions by estimating the values of neighboring pixels.
- **Time Series Forecasting:** Interpolation and curve fitting can be used to analyze and predict trends in time series data, such as stock prices, weather patterns, or sales figures.

In conclusion, interpolation and curve fitting are indispensable tools in the AI toolbox, allowing us to approximate functions and extract meaningful insights from data. By mastering these techniques, we can empower machines to learn from the world around them and make increasingly accurate predictions, ultimately driving the advancement of AI technology.

Numerical Integration and Differentiation

This section will delve into the fascinating world of numerical integration and differentiation, two essential mathematical techniques that play a crucial role in developing artificial intelligence (AI) systems. As we explore these methods, we will gain a deeper understanding of their underlying principles and appreciate their significance in the context of AI.

Numerical integration, also known as quadrature, is a method used to approximate the definite integral of a function. In AI, this technique is often employed to solve problems that involve continuous variables, such as calculating the area under a curve or determining the probability distribution of a random variable. Numerical integration is beneficial when dealing with difficult or impossible functions to integrate analytically.

There are several numerical integration techniques, each with its own set of advantages and limitations. Some of the most popular methods include the trapezoidal rule, Simpson's rule, and Gaussian quadrature. These techniques involve dividing the area under the curve into smaller sections, approximating the function within each section, and then summing the results to obtain the overall integral. The choice of method depends on factors such as the desired accuracy, computational efficiency, and the nature of the function being integrated.

On the other hand, numerical differentiation is the process of approximating the derivative of a function using discrete data points. In AI, this technique is often used to estimate the rate of change of a variable, which can be crucial for tasks such as optimization, control systems, and sensitivity analysis. Numerical differentiation is particularly valuable when dealing with difficult or impossible functions to differentiate analytically.

There are several numerical differentiation techniques, including finite difference methods, such as forward, backward, and central differences, as well as more advanced methods like Richardson extrapolation and automatic differentiation. These techniques involve estimating the slope of the function at a given point by considering the values of the function at nearby points. The choice of method depends on factors such as the desired accuracy, computational efficiency, and the nature of the function being differentiated.

In AI, numerical integration and differentiation are indispensable tools that enable us to tackle complex problems involving continuous variables. By approximating the integral and derivative of functions, we can gain valuable insights into the behavior of AI systems, optimize their performance, and, ultimately, create more intelligent and efficient solutions.

In conclusion, numerical integration and differentiation are vital mathematical techniques that significantly contribute to the development and success of AI systems. As we continue to push the boundaries of AI, it is essential for researchers and practitioners to have a solid grasp of these methods and their applications. By mastering these techniques, we can unlock AI's full potential and revolutionize how we live, work, and interact with the world around us.

Optimization Techniques in AI

In artificial intelligence (AI), optimization techniques play a crucial role in developing and improving algorithms and models. These methods are employed to find the best possible solution to a given problem, often by minimizing or maximizing a specific objective function. In this section, we will delve into the importance of optimization in AI, explore various optimization techniques, and discuss their applications in the field.

The Importance of Optimization in AI

Optimization is at the heart of many AI tasks, such as machine learning, neural networks, and natural language processing. Optimization techniques enable AI systems to learn from data, adapt to new inputs, and make accurate predictions by fine-tuning the parameters of a model. Furthermore, optimization methods can help reduce computational complexity and improve the efficiency of AI algorithms, making them more practical for real-world applications.

Gradient Descent and Its Variants

Gradient descent is a widely used optimization technique in AI, particularly in training neural networks. It is an iterative method that seeks to minimize a given objective function by updating the model's parameters in the direction of the negative gradient of the function. The learning rate, a hyperparameter that controls the step size of each update, plays a crucial role in the convergence of the algorithm.

Several variants of gradient descent, such as stochastic gradient descent (SGD) and mini-batch gradient descent, introduce randomness and batch processing, respectively, to improve the algorithm's efficiency and convergence properties. Additionally, adaptive gradient methods, such as AdaGrad, RMSProp, and Adam, have been developed to adjust the learning rate dynamically based on the history of gradients, further enhancing the optimization process.

Evolutionary Algorithms

Inspired by natural selection and evolution principles, evolutionary algorithms (EAs) are a family of optimization techniques that operate on a population of candidate solutions. EAs typically involve selection, crossover (recombination), and mutation to generate new solutions and evolve the population toward an optimal solution. Genetic algorithms, genetic programming, and differential evolution are popular examples of EAs used in AI applications.

Swarm Intelligence

Swarm intelligence is another class of optimization techniques that draws inspiration from the collective behavior of social organisms, such as ants, bees, and birds. These methods rely on the interaction of simple agents to explore the search space and converge toward an optimal solution. Particle swarm optimization (PSO) and ant colony optimization (ACO) are two well-known swarm intelligence techniques that have been successfully applied to various AI problems, including neural network training, clustering, and combinatorial optimization.

Other Optimization Techniques

Numerous other optimization techniques have been employed in AI, such as simulated annealing, tabu search, and hill climbing. These methods offer different trade-offs between exploration and exploitation, convergence speed, and computational complexity, making them suitable for different types of problems and constraints.

In conclusion, optimization techniques are indispensable tools in the development of AI systems, as they enable the fine-tuning of algorithms and models to achieve better performance and efficiency. By understanding and selecting the appropriate optimization method for a given problem, AI practitioners can harness the full potential of these techniques to advance the state of the art in artificial intelligence.

The Importance of Numerical Methods in AI Development

As we reach the end of this enlightening journey through the world of numerical methods, it is crucial to reflect on the significance of these mathematical techniques in artificial intelligence. The development of AI systems has been a groundbreaking endeavor, and the role of numerical methods in this process cannot be overstated. In this conclusion, we shall recapitulate the key points discussed in this chapter and emphasize the indispensable nature of numerical methods in AI development.

This chapter has explored various numerical techniques instrumental in solving equations and approximating functions. These methods, ranging from root-finding techniques to optimization algorithms, have proven invaluable tools for AI researchers and developers. By providing efficient and accurate solutions to complex mathematical problems, numerical methods have facilitated the creation of sophisticated AI models that can learn, adapt, and make decisions in myriad situations.

One of the most significant contributions of numerical methods to AI development is their ability to handle large-scale data and high-dimensional problems. As AI systems continue to evolve, they are required to process and analyze vast amounts of information. Numerical methods, such as interpolation and curve fitting, enable AI models to approximate functions and make predictions based on this data, enhancing their learning capabilities and overall performance.

Moreover, numerical integration and differentiation techniques have been crucial in developing AI algorithms that adapt and respond to dynamic environments. These methods allow AI systems to compute gradients and estimate the impact of various factors on their decision-making processes. Consequently, AI models can optimize their actions and make more informed choices, leading to improved outcomes and increased efficiency.

Optimization techniques, such as gradient descent and genetic algorithms, have also played a pivotal role in AI development. These methods enable AI systems to fine-tune their parameters and minimize errors, resulting in more accurate and reliable models. Furthermore, optimization techniques have facilitated the creation of AI models that tackle complex, real-world problems, such as natural language processing, image recognition, and autonomous navigation.

In conclusion, the importance of numerical methods in AI development is undeniable. These mathematical techniques have provided the foundation for AI algorithms and enabled the creation of advanced AI systems that can learn, adapt, and make decisions in an ever-changing world. As AI continues to progress and permeate various aspects of our lives, the role of numerical methods in shaping the future of this technology will remain paramount. By understanding and harnessing the power of these methods, we can unlock AI's full potential and revolutionize how we live, work, and interact with the world around us.

Chapter Summary

- Numerical methods are essential in AI development, providing the foundation for algorithms and models used in machine learning, data analysis, and optimization.
- Root-finding techniques, such as the bisection method and Newton-Raphson method, are crucial for solving equations and optimizing AI systems.
- Interpolation and curve fitting techniques, including linear interpolation and spline interpolation, enable AI models to approximate functions and make predictions based on limited data.
- Numerical integration and differentiation techniques, such as the trapezoidal rule and finite difference methods, are vital for analyzing and understanding the behavior of functions in AI systems.
- Optimization techniques, including gradient descent and evolutionary algorithms, are indispensable for fine-tuning AI models and improving their performance and efficiency.

- Numerical methods allow AI systems to handle large-scale data and high-dimensional problems, enhancing their learning capabilities and overall performance.
- AI models can optimize their actions and make more informed choices using numerical integration and differentiation techniques, leading to improved outcomes and increased efficiency.
- As AI continues to progress, the role of numerical methods in shaping the future of this technology will remain paramount, enabling the creation of advanced AI systems that can revolutionize how we live, work, and interact with the world around us.

7

OPTIMIZATION TECHNIQUES: ENHANCING AI PERFORMANCE

rtificial Intelligence (AI) has become integral to our daily lives, revolutionizing various industries and transforming how we interact with technology. From self-driving cars to personalized recommendations on streaming platforms, AI systems have proven their ability to learn and adapt to complex situations. However, the efficiency and effectiveness of these systems are heavily reliant on the optimization techniques employed during their development. This chapter will delve into the fascinating world of optimization techniques, exploring their significance in enhancing AI performance and their diverse applications.

Optimization techniques play a crucial role in developing AI systems, enabling fine-tuning algorithms and models to achieve the best possible performance. These techniques involve the process of adjusting the parameters and configurations of an AI system to minimize or maximize a specific objective function. The objective function, also known as the cost or loss function, is a mathematical representation of the system's performance, quantifying the difference between the predicted and actual output. By minimizing the value of the objective function, we can ensure that the AI system is making accurate predictions and decisions, thereby improving its overall performance.

Various optimization techniques are available, each with its unique approach to problem-solving and suitability for different types of AI systems. Some of the most widely used techniques include gradient descent and its variants, evolutionary algorithms, swarm intelligence, and reinforcement learning. These techniques have been instrumental in the success of AI systems across diverse domains, such as computer vision, natural language processing, robotics, and game playing.

The following sections will explore these optimization techniques in detail, discussing their underlying principles, advantages, and limitations. We will begin with gradient descent, the backbone of AI optimization, and its variants, which have been extensively used in training deep learning models. Next, we will delve into the world of evolutionary algorithms inspired by the process of natural selection and their applications in AI. Following that, we will examine swarm intelligence, a fascinating approach to collaborative problem-solving in AI inspired by the behavior of social insects. Finally, we will discuss reinforcement learning, a technique that teaches AI systems through trial and error, enabling them to learn from their interactions with the environment.

As we embark on this journey through optimization techniques, we hope to provide you with a comprehensive understanding of their significance in AI and inspire you to appreciate the intricate balance between mathematical rigor and creative problem-solving at the heart of AI performance optimization. With the rapid advancements in AI research and development, the future of AI performance optimization promises to be even more exciting and transformative, paving the way for more innovative, efficient, and human-like AI systems.

Gradient Descent and Its Variants: The Backbone of AI Optimization

In artificial intelligence (AI), optimization techniques play a crucial role in enhancing the performance of machine learning models. One of the most widely used and fundamental optimization algorithms is Gradient Descent. This powerful technique has been the backbone of AI optimization, enabling machines to learn and adapt to complex tasks efficiently. In this section, we will delve into the concept of Gradient Descent, its variants, and their significance in the world of AI.

Understanding Gradient Descent

Gradient Descent is an iterative optimization algorithm that aims to find the minimum value of a function, typically a cost or loss function, which measures the discrepancy between the predicted and actual outcomes. In simpler terms, Gradient Descent helps AI models minimize the error in their predictions by adjusting the model's parameters, such as weights and biases, in the direction of the steepest decrease in the cost function.

The core idea behind Gradient Descent is to compute the gradient (the vector of partial derivatives) of the cost function with respect to each parameter. The gradient represents the direction of the steepest increase in the function. By moving in the opposite direction of the gradient, we can iteratively update the parameters to minimize the cost function, ultimately reaching the optimal solution.

Variants of Gradient Descent

While the basic concept of Gradient Descent is straightforward, several variants of the algorithm cater to different scenarios and computational constraints. These variants primarily differ in how they update the model's parameters and how much data they use to compute the gradient. Let's explore the three main types of Gradient Descent:

- **Batch Gradient Descent:** This is the most basic form of Gradient Descent, where the gradient is calculated using the entire dataset. While this approach guarantees convergence to the global minimum for convex functions, it can be computationally expensive and slow for large datasets.
- **Stochastic Gradient Descent (SGD):** Unlike Batch Gradient Descent, SGD computes the gradient using only a single data point at each iteration. This makes the algorithm much faster and more suitable for large datasets. However, the trade-off is that the path to the optimal solution can be noisy and less direct due to the randomness introduced using a single data point.
- **Mini-batch Gradient Descent:** This variant balances Batch Gradient Descent and SGD by computing the gradient using a small subset of the dataset, called a mini-batch. Mini-batch Gradient Descent offers a compromise between computational efficiency and convergence stability, making it a popular choice in practice.

The Significance of Gradient Descent in AI

Gradient Descent and its variants have been instrumental in AI's success, particularly in deep learning. The algorithm's ability to efficiently navigate high-dimensional parameter spaces and find optimal solutions has made it the go-to optimization technique for training neural networks. Moreover, the flexibility offered by its variants allows practitioners to tailor the algorithm to their specific needs and computational resources.

In conclusion, Gradient Descent has undoubtedly earned its place as the backbone of AI optimization. Its simplicity, adaptability, and effectiveness have made it an indispensable tool in developing intelligent machines capable of tackling complex tasks. As AI continues to evolve, Gradient Descent and its variants will undoubtedly remain at the forefront of optimization techniques, driving the performance of AI models to new heights.

Evolutionary Algorithms: Harnessing the Power of Natural Selection

In artificial intelligence, optimization techniques play a crucial role in enhancing the performance of AI systems. One such technique that has gained significant attention in recent years is evolutionary algorithms. Drawing inspiration from the process of natural selection, these algorithms have proven to be highly effective in solving complex optimization problems. This section will delve into the fascinating world of evolutionary algorithms, exploring their underlying principles, various types, and applications in AI.

The Principles of Evolutionary Algorithms

Evolutionary algorithms are a family of optimization techniques that mimic the process of natural selection, the driving force behind the evolution of species. These algorithms operate on a population of potential solutions to a given problem, iteratively applying a set of biologically-inspired operators such as selection, crossover (recombination), and mutation. The population evolves over time through these operations, gradually converging towards an optimal or near-optimal solution.

The fundamental steps of an evolutionary algorithm can be summarized as follows:

I. **Initialization:** Generate an initial population of candidate solutions, typically at random.

2. **Evaluation:** Assess each individual's fitness in the population, i.e., how well it solves the problem at hand.
3. **Selection:** Choose a subset of individuals from the current population based on their fitness, favoring those with higher fitness values.
4. **Variation:** Apply crossover and mutation operators to the selected individuals, creating a new generation of offspring.
5. **Replacement:** Replace the current population with the newly generated offspring.
6. **Termination:** Repeat steps 2-5 until a stopping criterion is met, such as reaching a maximum number of generations or achieving a desired fitness level.

Types of Evolutionary Algorithms

There are several types of evolutionary algorithms, each with its unique characteristics and strengths. Some of the most widely used variants include:

- **Genetic Algorithms (GAs):** Perhaps the most well-known type of evolutionary algorithm, GAs employ binary strings to represent candidate solutions and use genetic operators such as selection, crossover, and mutation to evolve the population.
- **Genetic Programming (GP):** An extension of genetic algorithms, GP focuses on evolving computer programs or symbolic expressions to solve a given problem.
- **Evolutionary Strategies (ES):** Originating from the engineering optimization field, ES emphasizes using real-valued representations and self-adaptive mutation rates.
- **Differential Evolution (DE):** A population-based optimization technique that employs a unique mutation strategy, combining elements from multiple individuals to generate new offspring.

Applications of Evolutionary Algorithms in AI

The versatility and adaptability of evolutionary algorithms have made them a popular choice for tackling a wide range of optimization problems in AI. Some notable applications include:

- **Neural Network Training:** Evolutionary algorithms can optimize the weights and architecture of artificial neural networks, enhancing their performance in tasks such as pattern recognition, classification, and regression.
- **Feature Selection:** In machine learning, evolutionary algorithms can help identify the most relevant features for a given problem, reducing the dimensionality of the data and improving the efficiency of learning algorithms.
- **Game Playing:** Evolutionary algorithms have evolved strategies and heuristics for playing various games, from classic board games like chess and Go to modern video games.
- **Robotics:** In robotics, evolutionary algorithms have been used to optimize the design and control of robotic systems, enabling them to adapt to different environments and tasks.

In conclusion, evolutionary algorithms represent a powerful and versatile optimization technique that has found numerous applications in artificial intelligence. By harnessing the power of natural selection, these algorithms offer a unique approach to problem-solving, allowing AI systems to evolve and adapt in the face of complex and dynamic challenges. As AI advances, evolutionary algorithms will likely play an increasingly important role in shaping the future of AI performance optimization.

Swarm Intelligence: Collaborative Problem Solving in AI

In the vast and ever-evolving world of artificial intelligence, one of the most fascinating and effective optimization techniques is inspired by the natural world: swarm intelligence. This powerful approach to problem-solving is derived from the collective behavior of decentralized, self-organized systems, such as those observed in ant colonies, bird flocks, and fish schools. This section will delve into the captivating realm of swarm intelligence, exploring its origins, fundamental principles, and applications in AI.

The Inspiration: Nature's Collective Wisdom

The concept of swarm intelligence is rooted in observing nature's remarkable ability to solve complex problems through the collaborative efforts of simple agents. For instance, ants are individually limited in their cognitive abilities, yet as a colony, they can efficiently locate and transport food, build intricate nests, and defend against predators. Similarly, birds in a flock can navigate vast distances and avoid obstacles with remarkable precision, all without a central leader or explicit communication. These examples demonstrate the power of collective intelligence, where the whole is greater than the sum of its parts.

Principles of Swarm Intelligence

Swarm intelligence in AI is based on several key principles that mirror the behavior of natural swarms. These principles include:

- **Decentralization:** Swarm intelligence systems are composed of multiple agents that operate independently, without a central authority. This allows for greater flexibility, adaptability, and robustness in changing environments and problem parameters.
- **Local Interactions:** Agents in a swarm intelligence system rely on local information and interactions with their neighbors rather than global knowledge of the entire system. This enables the swarm to respond quickly to changes and efficiently distribute tasks among its members.

- **Emergence:** The collective behavior of a swarm intelligence system emerges from the simple, local interactions of its agents. This emergent behavior is often more sophisticated and effective than the behavior of any single agent, allowing the swarm to solve complex problems that would be difficult or impossible for an individual agent to tackle alone.

Applications of Swarm Intelligence in AI

Swarm intelligence has been successfully applied to various AI problems, including optimization, robotics, and data analysis. Some notable examples include:

- **Particle Swarm Optimization (PSO)** is a popular optimization algorithm inspired by the social behavior of bird flocks. In PSO, a swarm of particles moves through a search space, adjusting their positions based on their own experiences and those of their neighbors. This collaborative search process allows the swarm to locate optimal solutions to complex problems efficiently.
- **Ant Colony Optimization (ACO)** is another powerful optimization technique inspired by the foraging behavior of ants. In ACO, artificial ants traverse a graph representing a problem space, depositing pheromone trails that guide other ants toward promising solutions. Over time, the collective pheromone trails converge on the optimal solution, allowing the swarm to solve complex combinatorial problems, such as the traveling salesperson problem.
- **Swarm intelligence** principles have been applied to robotics, enabling the development of multi-robot systems that can collaboratively perform tasks such as exploration, mapping, and search and rescue. These swarm robotic systems are highly robust and adaptable, as they can

continue to function even if individual robots fail or are removed from the swarm.

In conclusion, swarm intelligence offers a captivating and powerful approach to problem-solving in AI, drawing inspiration from the collective wisdom of nature's swarms. By harnessing the principles of decentralization, local interactions, and emergence, swarm intelligence techniques have demonstrated remarkable success in a wide range of AI applications, from optimization to robotics. As we continue to explore and develop new swarm intelligence algorithms, we expect to see even greater advancements in AI performance optimization, further unlocking the potential of artificial intelligence to tackle the most complex and challenging problems of our time.

Reinforcement Learning: Teaching AI through Trial and Error

In the ever-evolving landscape of artificial intelligence, one of the most promising and exciting areas of research is reinforcement learning (RL). This powerful optimization technique, inspired by how humans and animals learn from their experiences, can potentially revolutionize AI performance. In this section, we will delve into the fascinating world of reinforcement learning, exploring its core concepts, applications, and the challenges it presents.

At its core, reinforcement learning is a trial-and-error approach to problem-solving. An AI agent, or learner, interacts with its environment, taking action and receiving feedback through rewards or penalties. The agent aims to learn an optimal policy, a set of rules that dictate the best action to take in each situation to maximize its cumulative reward over time. This process of exploration and exploitation allows the AI to adapt and improve its performance, even in complex and uncertain environments.

One of the critical components of reinforcement learning is the concept of the Markov Decision Process (MDP). MDPs provide a mathematical framework for modeling decision-making when the outcome is uncertain and depends on the current state and the chosen action.

The AI agent can determine the optimal policy and improve its decision-making capabilities by solving the MDP.

Several algorithms and approaches are used in reinforcement learning, including value iteration, policy iteration, Q-learning, and deep Q-networks (DQNs). Each of these methods has its strengths and weaknesses, and the choice of algorithm depends on the specific problem and the desired level of performance.

Reinforcement learning has found numerous applications in various fields, such as robotics, finance, healthcare, and gaming. For instance, Google's DeepMind developed AlphaGo. This AI program defeated the world champion in the ancient board game of Go, using a combination of deep neural networks and reinforcement learning. Similarly, RL has been used to optimize trading strategies in financial markets, develop adaptive treatment plans for patients with chronic conditions, and even teach drones to navigate complex environments autonomously.

Despite its immense potential, reinforcement learning also presents several challenges. One of the primary issues is the trade-off between exploration and exploitation. The AI agent must balance the need to explore new actions and states to learn more about the environment with the need to exploit its current knowledge to maximize rewards. This balance is crucial for achieving optimal performance, but finding the right balance can be a complex and computationally expensive task.

Another challenge is the so-called "curse of dimensionality," which arises when the AI agent must deal with a large number of states and actions. As the size of the state-action space increases, the computational complexity of reinforcement learning algorithms grows exponentially, making it difficult to find optimal solutions in a reasonable amount of time.

In conclusion, reinforcement learning is a powerful optimization technique with great promise for enhancing AI performance across various applications. By harnessing the power of trial and error, AI agents can learn to navigate complex and uncertain environments, adapting their behavior to achieve optimal results. As research in this field advances, we can expect to see even more impressive achievements and breakthroughs in artificial intelligence.

The Future of AI Performance Optimization

As we have explored throughout this chapter, optimization techniques play a crucial role in enhancing artificial intelligence systems' performance. From the fundamental gradient descent algorithm to the more advanced evolutionary algorithms, swarm intelligence, and reinforcement learning, these methods have significantly contributed to the development and success of AI applications across various domains.

Looking ahead, AI performance optimization's future is promising and challenging. With the rapid advancements in technology and the ever-growing demand for more efficient and intelligent systems, researchers and practitioners are continuously seeking innovative ways to improve the performance of AI algorithms. In this pursuit, we expect to witness the emergence of new optimization techniques and the refinement of existing ones.

One potential area of growth lies in the integration of different optimization methods. By combining the strengths of various techniques, such as the adaptability of evolutionary algorithms and the collaborative problem-solving capabilities of swarm intelligence, we can develop hybrid optimization approaches to tackle complex problems more effectively.

Moreover, as AI systems become increasingly sophisticated, the need for optimization techniques to handle large-scale, high-dimensional problems will become more pressing. This may lead to developing novel algorithms that can efficiently explore vast search spaces and promptly identify optimal solutions.

Another exciting prospect for the future of AI performance optimization is incorporating human intuition and expertise into the optimization process. By leveraging the unique problem-solving abilities of humans, AI systems can overcome some of the limitations of traditional optimization techniques and achieve even greater performance levels.

Furthermore, as AI continues to permeate various aspects of our lives, ethical considerations surrounding optimizing AI systems will become increasingly important. Ensuring that AI algorithms are optimized fairly, transparently, and respect society's values will be a critical challenge for researchers and practitioners alike.

In conclusion, the future of AI performance optimization is filled with opportunities and challenges. As we continue to push the boundaries of what AI systems can achieve, developing and refining optimization techniques will remain an essential component of this journey. By embracing the spirit of innovation and collaboration, we can look forward to a future where AI systems are more efficient, effective, and aligned with the values and needs of our diverse world.

Chapter Summary

- Optimization techniques are crucial for enhancing AI performance, as they enable fine-tuning algorithms and models to achieve the best possible results.
- Gradient descent and its variants, such as batch, stochastic, and mini-batch gradient descent, are widely used optimization techniques in AI, particularly deep learning.
- Evolutionary algorithms, inspired by the process of natural selection, offer a unique approach to problem-solving and have been applied to various AI domains, including neural network training and feature selection.
- Swarm intelligence, derived from the collective behavior of decentralized, self-organized systems in nature, has been successfully applied to AI problems such as particle swarm optimization and ant colony optimization.
- Reinforcement learning, a trial-and-error approach to problem-solving, has shown great potential in various fields, including robotics, finance, healthcare, and gaming.
- The future of AI performance optimization may involve integrating different optimization methods, developing

novel algorithms for large-scale problems, and incorporating human intuition and expertise.

- Ethical considerations surrounding the optimization of AI systems, such as fairness and transparency, will become increasingly important as AI continues to permeate various aspects of our lives.
- By embracing innovation and collaboration, we can look forward to a future where AI systems are more efficient, effective, and aligned with the values and needs of our diverse world.

8

GAME THEORY: ANALYZING STRATEGIC DECISION-MAKING

G ame theory, a branch of mathematics and economics, has emerged as a powerful tool for understanding and predicting the behavior of intelligent agents in various situations. It is a study of strategic decision-making, where multiple players interact and make decisions based on their objectives and the actions of others. In artificial intelligence (AI), game theory is crucial in designing algorithms and systems that can adapt and respond to the complex and dynamic environments they encounter.

The importance of game theory in AI stems from its ability to model and analyze the interactions between multiple agents, whether human or artificial. As AI systems become more advanced and integrated into our daily lives, they must be able to navigate and make decisions in environments that involve other intelligent agents. Game theory provides a framework for understanding these interactions and designing AI systems that can make optimal decisions in the face of uncertainty and competition.

This chapter will explore the fundamentals of game theory and its applications in AI. We will begin by introducing key concepts and terminology used in game theory, such as players, strategies, payoffs, and equilibria. Next, we will delve into the different types of games and their relevance to AI, including zero-sum games, cooperative games, and non-cooperative games. We will then discuss various techniques and algorithms for solving game theory problems, focusing on their implementation in AI systems.

To illustrate the practical applications of game theory in AI and machine learning, we will present real-world examples that showcase how game theory has improved decision-making in areas such as robotics, natural language processing, and autonomous vehicles. Finally, we will conclude by discussing the future of game theory in AI and its potential impact on strategic decision-making in the rapidly evolving world of artificial intelligence.

By the end of this chapter, readers will have a solid understanding of game theory and its significance in developing AI systems. They will also gain insights into the various techniques and algorithms used to solve game theory problems and how they can be applied to create intelligent agents capable of making strategic decisions in complex environments.

Key Concepts and Terminology in Game Theory

In this section, we will delve into the essential concepts and terminology that form the foundation of game theory. Understanding these fundamental ideas, you will be better equipped to analyze strategic decision-making in artificial intelligence (AI) and other fields.

Players and Strategies

At the core of any game are the players, who are the decision-makers in the game. Players can be algorithms, agents, or even human beings in AI. Each player has a set of possible actions or decisions, known as strategies, that they can choose from to achieve their objectives. Strategies can be simple, like choosing a number in a lottery, or complex, like deciding the next move in a chess game.

Payoffs and Utility

The outcome of a game depends on the strategies chosen by the players. Each player prefers the possible outcomes represented by a payoff or utility function. The utility function assigns a numerical value to each outcome, reflecting the player's satisfaction or benefit from that outcome. In AI, the utility function can represent an algorithm's performance or an agent's success in achieving its goal.

Nash Equilibrium

A central concept in game theory is the Nash equilibrium, named after the mathematician John Nash. A Nash equilibrium is a stable state in which no player can improve their payoff by unilaterally changing their strategy, given the other players' strategies. In other words, it is a situation where each player's strategy is the best response to the other players' strategies. The existence of a Nash equilibrium in a game is a powerful tool for predicting the outcome of strategic interactions.

Zero-Sum and Non-Zero-Sum Games

Games can be classified as zero-sum or non-zero-sum based on the relationship between the players' payoffs. In a zero-sum game, the total payoff for all players is constant, meaning that one player's gain is another player's loss. Classic examples of zero-sum games include poker and chess. In contrast, non-zero-sum games allow for the possibility of mutual gain or loss, such as in cooperative games or negotiation scenarios.

Perfect and Imperfect Information

Another important distinction in game theory is between games of perfect and imperfect information. In a game of perfect information, all players have complete knowledge of the game's structure, including the strategies and payoffs of the other players. Chess is an example of a game with perfect information. In games of imperfect information, players have incomplete knowledge about the game, such as the strategies or payoffs of other players. Poker is an example of a game with imperfect information, as players do not know the cards held by their opponents.

By understanding these key concepts and terminology in game theory, you will be better prepared to analyze strategic decision-making in AI and other fields. In the next section, we will explore the different types of games and their applications in AI, further expanding your knowledge of this fascinating subject.

Solving Game Theory Problems: Techniques and Algorithms

This section will delve into the various techniques and algorithms used to solve game theory problems. These methods are crucial for understanding strategic decision-making in AI and machine learning applications. By mastering these techniques, AI systems can make more informed decisions and better predict the actions of other agents in a given environment.

Dominant Strategy Equilibrium

A dominant strategy equilibrium occurs when a player has a strategy that yields the best outcome, regardless of the strategies chosen by other players. In such cases, the player will always choose this dominant strategy. To identify a dominant strategy equilibrium, one must analyze each player's payoff and determine if a particular strategy consistently yields the highest payoff.

Nash Equilibrium

Named after the mathematician John Nash, a Nash equilibrium is a set of strategies where no player can improve their payoff by unilaterally changing their strategy. In other words, each player's strategy is the best response to the strategies of all other players. To find a Nash equilibrium, one can use various algorithms, such as the best-response algorithm or the Lemke-Howson algorithm.

Minimax Algorithm

The minimax algorithm is a decision-making technique used in two-player, zero-sum games. It involves each player minimizing the maximum possible loss they could incur. The algorithm works by recursively evaluating the game tree, assigning a value to each node based on the best possible outcome for the player. The minimax algorithm is handy in AI applications, such as board games like chess or tic-tac-toe, where the AI must make decisions based on the possible moves of its opponent.

Alpha-Beta Pruning

Alpha-beta pruning is an optimization technique used in conjunction with the minimax algorithm. It reduces the number of nodes that need to be evaluated in the game tree by eliminating branches that do not contribute to the final decision. This allows the AI to search deeper into the game tree and make more informed decisions while reducing computational time.

Reinforcement Learning

Reinforcement learning is a type of machine learning where an AI agent learns to make decisions by interacting with its environment and receiving feedback in the form of rewards or penalties. In game theory, reinforcement learning can be used to train AI agents to find optimal strategies in various games. Techniques like Q-learning and deep reinforcement learning have been successfully applied to complex games like Go and poker, allowing AI systems to outperform human players.

In conclusion, solving game theory problems is essential for AI systems to make strategic decisions and predict the actions of other agents. By mastering techniques such as dominant strategy equilibrium, Nash equilibrium, minimax algorithm, alpha-beta pruning, and reinforcement learning, AI systems can become more effective in a wide range of applications, from board games to complex real-world scenarios. As AI advances, we can expect to see even more sophisticated game theory techniques and algorithms being developed and applied to strategic decision-making.

Real-World Examples of Game Theory in AI and Machine Learning

In this section, we will delve into the fascinating world of real-world applications of game theory in artificial intelligence (AI) and machine learning. By examining these examples, we can better understand how game theory principles are employed to solve complex problems and make strategic decisions in various domains.

Autonomous Vehicles and Traffic Management

One of the most promising applications of game theory in AI is in the realm of autonomous vehicles. As self-driving cars become more prevalent, traffic management systems must adapt to accommodate these vehicles' unique characteristics. Game theory can be used to model the interactions between autonomous vehicles and human-driven cars, allowing for the development of optimal traffic management strategies that minimize congestion and maximize safety.

For instance, researchers have used game theory to design algorithms that enable autonomous vehicles to negotiate intersections without traffic signals. By treating each vehicle as a player in a game, the vehicles can make strategic decisions about when to enter the intersection, taking into account the actions of other vehicles and minimizing the risk of collisions.

Multi-Agent Systems and Robotics

Game theory is also a valuable tool in the design and analysis of multi-agent systems, where multiple AI agents interact with each other to achieve a common goal. In robotics, for example, game theory can be used to model the interactions between multiple robots working together to complete a task, such as assembling a structure or exploring an unknown environment.

By applying game theory principles, researchers can develop algorithms enabling robots to cooperate effectively, allocate resources efficiently, and avoid conflicts. This can lead to more robust and resilient multi-agent systems adapting to changing conditions and achieving their objectives more efficiently.

Auctions and Market Design

AI and machine learning have been increasingly employed in designing and analyzing auctions and markets. Game theory plays a crucial role in understanding the strategic behavior of participants in these settings, allowing for the development of more efficient and fair mechanisms.

For example, AI-powered algorithms have been used to design auctions for allocating radio spectrum licenses, ensuring that the licenses are allocated to the parties that value them the most while generating revenue for the government. Similarly, game theory has been applied to the design of online advertising markets, where advertisers bid for the opportunity to display their ads to users.

Cybersecurity and Network Defense

In cybersecurity, game theory has emerged as a powerful tool for modeling the interactions between attackers and defenders. By treating cybersecurity as a game, researchers can develop strategies for defending networks and systems against various types of attacks, taking into account the actions and motivations of the attackers.

For instance, machine learning algorithms can predict attackers' behavior based on historical data, allowing defenders to allocate resources more effectively and prioritize their efforts. Game theory can also be used to design incentives for users to adopt secure practices, such as using strong passwords and keeping their software up-to-date.

Social Network Analysis and Influence Maximization

Finally, game theory has found applications in analyzing social networks and designing strategies for maximizing influence in these networks. By modeling the interactions between individuals in a social network as a game, researchers can develop algorithms that identify the most influential nodes and devise strategies for spreading information or promoting products and services.

Machine learning techniques can be employed to analyze large-scale social network data, allowing for identifying patterns and trends that can inform the design of influence maximization strategies. This has applications in areas such as viral marketing, political campaigns, and public health interventions.

In conclusion, the real-world examples of game theory in AI and machine learning demonstrate the versatility and power of this mathematical framework in addressing complex problems and enabling strategic decision-making. As AI continues to advance and permeate various domains, we can expect game theory to play an increasingly important role in shaping the future of AI-driven systems and applications.

The Future of Game Theory in AI and Strategic Decision-Making

As we have explored throughout this chapter, game theory plays a crucial role in developing and advancing artificial intelligence. By providing a mathematical framework for analyzing strategic decision-making, game theory enables AI systems to make more informed choices and adapt to complex, dynamic environments. In this concluding section, we will discuss the future of game theory in AI and its potential impact on strategic decision-making.

Multi-agent systems are one of the most promising areas for applying game theory in AI. These systems consist of multiple autonomous agents that interact with each other to achieve specific goals. By incorporating game theory into the design of multi-agent systems, researchers can develop AI agents capable of cooperating, competing, and negotiating with one another in a wide range of scenarios. This could lead to more efficient and robust AI systems that tackle complex problems in transportation, logistics, and resource allocation.

Another exciting avenue for the future of game theory in AI is the development of new algorithms and techniques for solving game-theoretic problems. As AI systems become more sophisticated, they will need to be able to handle larger and more complex games. This will require the development of new algorithms that can efficiently compute optimal strategies and equilibria in these games. Researchers are already making progress in this area, with recent advances in techniques such as deep reinforcement learning and neural network-based approaches showing great promise for solving game-theoretic problems.

In addition to these technical advancements, the future of game theory in AI will also be shaped by its integration with other disciplines and methodologies. For example, researchers are increasingly exploring the connections between game theory and fields such as economics, psychology, and sociology. By combining insights from these disciplines with game-theoretic models, AI systems can be designed to understand better and predict human behavior, leading to more effective and ethical decision-making.

Finally, as AI continues to permeate various aspects of our lives, understanding and addressing the ethical implications of AI systems that utilize game theory cannot be overstated. As AI systems become more capable of making strategic decisions, ensuring that these decisions align with our values and ethical principles is essential. This will require ongoing collaboration between AI researchers, ethicists, policymakers, and other stakeholders to develop guidelines and best practices for the responsible use of game theory in AI.

In conclusion, the future of game theory in AI and strategic decision-making is bright, with numerous opportunities for innovation and growth. By continuing to explore the applications of game theory in AI, developing new algorithms and techniques, and addressing the ethical implications of AI systems that utilize game theory, we can unlock the full potential of this powerful mathematical framework and pave the way for a new era of intelligent, strategic decision-making.

Chapter Summary

- Game theory is a powerful tool for understanding and predicting the behavior of intelligent agents in various situations, making it crucial for designing AI algorithms and systems that can adapt to complex environments.
- Key concepts in game theory include players, strategies, payoffs, Nash equilibrium, zero-sum and non-zero-sum games, and perfect and imperfect information.
- Different types of games, such as zero-sum, non-zero-sum, cooperative and non-cooperative, and stochastic and deterministic games, have various applications in AI and can help develop intelligent agents capable of strategic decision-making.

- Techniques and algorithms for solving game theory problems include dominant strategy equilibrium, Nash equilibrium, minimax algorithm, alpha-beta pruning, and reinforcement learning.
- Game theory has real-world applications in AI and machine learning, such as autonomous vehicles and traffic management, multi-agent systems and robotics, auctions, and market design, cybersecurity and network defense, and social network analysis and influence maximization.
- The future of game theory in AI lies in developing multi-agent systems, new algorithms and techniques for solving game-theoretic problems, and integration with other disciplines such as economics, psychology, and sociology.
- Addressing the ethical implications of AI systems that utilize game theory is essential to ensure that strategic decisions align with our values and ethical principles, requiring collaboration between AI researchers, ethicists, policymakers, and other stakeholders.
- By exploring the applications of game theory in AI, developing new algorithms and techniques, and addressing ethical implications, we can unlock the full potential of game theory and pave the way for a new era of intelligent, strategic decision-making.

9

INFORMATION THEORY: QUANTIFYING AND ENCODING DATA

I n the world of artificial intelligence (AI), the ability to process, analyze, and understand vast amounts of data is crucial. As AI systems become more sophisticated, they require efficient methods to manage and utilize this data effectively. This is where Information Theory comes into play. In this chapter, we will delve into the fascinating realm of Information Theory and explore its significance in the development and advancement of AI.

Information Theory, a branch of mathematics and computer science, was introduced by Claude Shannon in his groundbreaking 1948 paper, "A Mathematical Theory of Communication." At its core, Information Theory is concerned with quantifying, encoding, and transmitting data in the most efficient and reliable manner possible. It provides a mathematical framework for understanding the fundamental limits of communication systems and offers valuable insights into the design and optimization of these systems.

In the context of AI, Information Theory plays a pivotal role in various aspects of data processing and analysis. From data compression techniques that enable efficient storage and transmission of information to error detection and correction mechanisms that ensure data integrity, Information Theory serves as a foundation for many AI applications.

One of the key concepts in Information Theory is entropy, which measures the uncertainty or randomness of data. In AI, understanding entropy is essential for tasks such as feature selection, model complexity, and decision-making processes. AI systems can make more informed decisions and improve their overall performance by quantifying data uncertainty.

Moreover, Information Theory is instrumental in developing machine learning algorithms, particularly in unsupervised learning and reinforcement learning. These algorithms rely on the principles of Information Theory to discover patterns, extract meaningful features, and optimize their learning processes.

As AI advances, the importance of Information Theory cannot be overstated. By providing a solid foundation for understanding the intricacies of data and communication systems, Information Theory enables AI researchers and practitioners to push the boundaries of what is possible in the field. In the following sections, we will delve deeper into the key concepts of Information Theory, such as entropy, data compression, error detection and correction, and channel capacity, and explore their applications in AI.

Entropy: Measuring the Uncertainty of Data

In the realm of artificial intelligence, the ability to quantify and manage the uncertainty of data is of paramount importance. This is where the concept of entropy comes into play. Entropy, a term borrowed from thermodynamics, is a measure of the uncertainty or randomness of a set of data. This section will delve into the significance of entropy, its mathematical representation, and its applications in AI.

To begin with, let us understand why entropy is crucial in the context of AI. Artificial intelligence systems often deal with vast amounts of data, which may contain varying degrees of uncertainty. By quantifying this uncertainty, AI algorithms can make better-informed decisions, optimize their performance, and reduce the risk of errors. In essence, entropy serves as a guiding force for AI systems to navigate through the complex landscape of data.

Now, let us explore the mathematical representation of entropy. Entropy is denoted by the letter 'H' and is calculated using the following formula:

$$H(X) = - \sum P(x) * log2(P(x))$$

Here, 'X' represents a discrete random variable with possible outcomes $x1$, $x2$, ..., xn, and P(x) denotes the probability of each outcome. The logarithm is base 2, meaning entropy is measured in bits. In simpler terms, entropy quantifies the average amount of information required to describe the outcome of a random variable.

To illustrate this concept, let us consider a simple example. Suppose we have a fair coin with equal probabilities of landing heads (H) or tails (T). The entropy of this coin can be calculated as follows:

$$H(Coin) = - [P(H) * log2(P(H)) + P(T) * log2(P(T))]$$
$$= - [(0.5 * log2(0.5)) + (0.5 * log2(0.5))]$$
$$= 1 \ bit$$

This result indicates that, on average, we need 1 bit of information to describe the outcome of a fair coin toss. In contrast, if we had a biased coin with a 90% chance of landing heads, the entropy would be lower, reflecting the reduced uncertainty in the outcome.

In AI, entropy is vital in various applications, such as decision trees, data compression, and natural language processing. For instance, in decision tree algorithms, entropy determines the best attribute for splitting the data at each node, optimizing the classification process. Similarly, in data compression techniques, entropy helps identify the most efficient encoding schemes to minimize storage and transmission costs.

In conclusion, entropy is a powerful tool for measuring data uncertainty in artificial intelligence systems. By quantifying this uncertainty, AI algorithms can optimize their performance, make better-informed decisions, and ultimately contribute to advancing the field. As we continue to explore the fascinating world of information theory, we will uncover more ways in which entropy and other related concepts can help shape the future of AI.

Data Compression Techniques: Efficient Storage and Transmission

In the realm of artificial intelligence, the ability to store and transmit data efficiently is of paramount importance. As AI systems grow in complexity and the volume of data they process increases, the need for effective data compression techniques becomes increasingly crucial. This section will explore various data compression methods that enable AI systems to store and transmit data more efficiently, enhancing their overall performance.

Data compression refers to the process of reducing the size of a data file without compromising its original content. This is achieved by identifying and eliminating redundancies within the data, allowing it to be represented in a more compact form. There are two primary types of data compression: lossless and lossy.

Lossless Compression

Lossless compression techniques ensure that the original data can be perfectly reconstructed from the compressed data. This is particularly important in AI applications where data integrity is critical, such as in medical imaging or financial analysis. Some common lossless compression algorithms include:

- **Huffman Coding:** This technique assigns shorter binary codes to more frequently occurring data elements, resulting in a reduced overall file size. Huffman coding is widely used in text compression and forms the basis of several popular file formats, such as ZIP and GZIP.
- **Run-Length Encoding (RLE):** RLE is a simple compression method that replaces consecutive occurrences of the same data element with a single instance of the element, followed by the number of repetitions. This technique is particularly effective for compressing data with large areas of uniformity, such as images with solid color backgrounds.
- **Lempel-Ziv-Welch (LZW) Algorithm:** LZW is a dictionary-based compression algorithm that replaces repeated occurrences of data with references to a dictionary entry. This method is employed in various file formats, including GIF and TIFF images.

Lossy Compression

On the other hand, Lossy compression techniques sacrifice some degree of data fidelity in exchange for greater compression efficiency. These methods are typically employed in situations where a small loss of data quality is acceptable, such as in audio or video compression. Some widely used lossy compression algorithms include:

- **JPEG (Joint Photographic Experts Group):** JPEG is a popular image compression standard that utilizes discrete cosine transform (DCT) and quantization to reduce the size of image files. By adjusting the level of compression, users can balance the trade-off between file size and image quality.

- **MP3 (MPEG-1 Audio Layer III):** MP3 is a widely used audio compression format that employs perceptual coding techniques to remove audio data that is less perceptible to the human ear. This allows for significant reductions in file size while maintaining an acceptable level of audio quality.
- **H.264/AVC (Advanced Video Coding):** H.264 is a video compression standard that utilizes motion compensation, spatial prediction, and other advanced techniques to achieve high compression efficiency while maintaining good video quality. It is widely used in video streaming and broadcasting applications.

In conclusion, data compression techniques play a vital role in efficiently storing and transmitting data in AI systems. By employing a combination of lossless and lossy compression methods, AI developers can optimize their systems to handle vast amounts of data without sacrificing performance or integrity. As AI advances, developing new and improved data compression techniques will remain an essential area of research and innovation.

Error Detection and Correction: Ensuring Data Integrity

In the realm of artificial intelligence, the accuracy and reliability of data are of paramount importance. As AI systems process vast amounts of information, even the smallest errors can have significant consequences. This is where error detection and correction techniques come into play, ensuring the integrity of data as it is transmitted and stored. In this section, we will delve into the fundamental concepts of error detection and correction and explore how these methods contribute to AI systems' overall efficiency and effectiveness.

The Need for Error Detection and Correction

As data travels through various channels, it is susceptible to noise, interference, and other factors that can introduce errors. These errors manifest as flipped bits, missing data, or corrupted information. To maintain the integrity of the data, it is crucial to identify and correct these errors as they occur. Error detection and correction techniques are designed to address this need, providing a means to identify and rectify inaccuracies in the data.

Error Detection Techniques

Error detection techniques are the first line of defense in ensuring data integrity. These methods involve adding redundant information to the original data, which can be used to detect errors upon receipt. Some common error detection techniques include:

- **Parity Bits:** A parity bit is a binary digit added to a group of bits, ensuring that the total number of 1s in the group is either even (even parity) or odd (odd parity). This simple technique can detect single-bit errors but is not effective for multiple-bit errors.
- **Checksums:** A checksum is a value calculated from the original data transmitted alongside the data. Upon receipt, the receiver calculates the checksum from the received data and compares it to the transmitted checksum. If the values match, the data is assumed to be error-free.
- **Cyclic Redundancy Check (CRC):** CRC is a more advanced error detection technique that involves generating a polynomial code from the original data. This code is then transmitted alongside the data, and the receiver uses it to verify the integrity of the received information.

Error Correction Techniques

While error detection techniques can identify errors, they do not provide a means to correct them. On the other hand, error correction techniques detect errors and enable the receiver to reconstruct the original data. Some common error correction techniques include:

The **Hamming code** is a linear block code that adds redundant bits to the original data, allowing the receiver to detect and correct single-bit errors. This technique is widely used in computer memory systems and other applications where single-bit errors are common.

The **Reed-Solomon code** is a more advanced error correction technique that can correct multiple errors in a data block. This method is particularly useful in communication systems, where noise and interference can cause multiple errors in a single transmission.

Turbo Codes and LDPC Codes are modern error correction techniques that employ iterative decoding algorithms to achieve near-optimal error correction performance. They are widely used in applications such as deep-space communication and wireless communication systems.

In conclusion, error detection and correction techniques are vital in ensuring data integrity in AI systems. By identifying and rectifying errors, these methods contribute to AI algorithms' overall efficiency and effectiveness, enabling them to make accurate predictions and decisions based on reliable information. As AI advances, the importance of robust error detection and correction mechanisms will only grow, paving the way for even more sophisticated and reliable AI systems.

Channel Capacity and the Noisy Channel Coding Theorem

Artificial intelligence's ability to transmit and process information accurately and efficiently is of utmost importance. As we delve into the fascinating world of information theory, we must explore the concept of channel capacity and the noisy channel coding theorem, which plays a crucial role in understanding the limits and potential of communication systems.

Channel capacity refers to the maximum rate at which information can be transmitted over a communication channel without losing accuracy due to noise or interference. It is measured in bits per second (bps) and serves as a benchmark for the performance of communication systems. In AI, channel capacity is vital as it determines the speed and reliability of data transmission between various components of a system.

The noisy channel coding theorem, a cornerstone of information theory, was introduced by Claude Shannon in 1948. This theorem addresses the fundamental question of how to transmit information reliably over a noisy channel, which is a communication channel that introduces errors or distortions in the transmitted data. The theorem states that, given a noisy channel with a certain capacity, it is possible to transmit information at a rate close to this capacity with an arbitrarily low probability of error, provided that an appropriate coding scheme is used.

Let's consider a simple example to understand better the noisy channel coding theorem. Imagine you are trying to send a message to a friend using a walkie-talkie. However, the signal is weak, and there is a lot of static noise. You may repeat the message several times to ensure that your friend receives the message accurately. This repetition is a form of error-correcting code, which increases the reliability of the transmission at the cost of a reduced effective data rate.

In the context of AI, the noisy channel coding theorem has significant implications. It highlights the importance of developing efficient coding schemes that maximize the information transmission rate while minimizing the probability of errors. This is particularly relevant in applications such as speech recognition, natural language processing, and computer vision, where the input data is often noisy or incomplete.

Moreover, the noisy channel coding theorem also provides a theoretical foundation for understanding the limits of communication systems. By quantifying the trade-off between data rate and error probability, it offers valuable insights into designing robust and efficient AI systems that can operate in real-world conditions with varying noise and interference levels.

In conclusion, channel capacity and the noisy channel coding theorem are essential concepts in information theory that profoundly impact the field of artificial intelligence. By understanding the limits and potential of communication systems, researchers and engineers can develop innovative AI solutions that can effectively process and transmit information in a noisy and uncertain world.

The Role of Information Theory in Advancing AI

As we reach the end of our journey through the fascinating world of Information Theory, it is essential to reflect on its significant role in advancing Artificial Intelligence. Throughout this chapter, we have explored various concepts and techniques that form the foundation of Information Theory and contribute to the development and optimization of AI systems.

The importance of Information Theory in AI cannot be overstated. It provides a robust framework for quantifying and encoding data, which is the lifeblood of any AI system. By understanding data uncertainty through entropy, we can develop more efficient algorithms to process and analyze vast amounts of information. This, in turn, allows AI systems to make better predictions and decisions, ultimately enhancing their performance.

Data compression techniques, as discussed in this chapter, are crucial for efficiently storing and transmitting information. As AI systems grow in complexity and the amount of data they process increases exponentially, the need for effective data compression becomes even more critical. By employing these techniques, we can reduce AI systems' storage and bandwidth requirements, making them more accessible and cost-effective.

Error detection and correction methods are vital in ensuring data integrity, which is paramount for the reliability and accuracy of AI systems. By identifying and rectifying errors in data transmission, we can maintain the quality of information being fed into AI algorithms. This improves the overall performance of AI systems and helps build trust in their outputs.

The concept of channel capacity and the Noisy Channel Coding Theorem provides valuable insights into data transmission limits and the trade-offs between speed, reliability, and efficiency. By understanding these limits, AI researchers and engineers can design better communication protocols and optimize the performance of AI systems in real-world scenarios.

In conclusion, Information Theory is an indispensable tool in the ever-evolving field of Artificial Intelligence. Its principles and techniques profoundly impact how we design, develop, and optimize AI systems. As we continue to push the boundaries of AI, Information Theory will undoubtedly remain a cornerstone in our quest to create intelligent machines that can learn, adapt, and thrive in an increasingly complex and data-driven world.

Chapter Summary

- Information Theory, introduced by Claude Shannon, is a branch of mathematics and computer science that focuses on quantifying, encoding, and transmitting data efficiently and reliably, making it crucial for AI systems.
- Entropy is a key concept in Information Theory, measuring the uncertainty or randomness of data. Understanding entropy is essential for AI tasks such as feature selection, model complexity, and decision-making processes.
- Data compression techniques, including lossless and lossy compression, are vital for efficiently storing and transmitting information in AI systems, enabling them to handle vast amounts of data without sacrificing performance or data integrity.
- Error detection and correction techniques are crucial in ensuring data integrity in AI systems and identifying and rectifying errors to maintain the quality of information fed into AI algorithms.
- Channel capacity refers to the maximum rate at which information can be transmitted over a communication channel without losing accuracy due to noise or

interference, determining the speed and reliability of data
transmission in AI systems.

- The Noisy Channel Coding Theorem, introduced by
Shannon, states that transmitting information at a rate close
to the channel capacity with an arbitrarily low probability of
error is possible, provided that an appropriate coding
scheme is used.

- Information Theory provides a theoretical foundation for
understanding the limits of communication systems,
offering valuable insights into designing robust and efficient
AI systems that can operate in real-world conditions with
varying levels of noise and interference.

- As AI continues to advance, Information Theory will remain
an indispensable tool in developing and optimizing AI
systems, enabling researchers and engineers to create
intelligent machines that can learn, adapt, and thrive in an
increasingly complex and data-driven world.

10

TOPOLOGY AND GEOMETRY: UNCOVERING HIDDEN STRUCTURES

I n artificial intelligence (AI), the quest for understanding and replicating human intelligence has led researchers to explore various mathematical concepts and techniques. Topology and geometry have emerged as essential tools in developing advanced AI algorithms and applications. This chapter aims to provide an insightful overview of the significance of topology and geometry in AI, delving into their foundational concepts, applications, and future prospects.

Topology, a branch of mathematics concerned with the study of spatial properties that are preserved under continuous transformations, has proven invaluable in analyzing complex data structures. Topology allows AI researchers to uncover hidden structures and patterns that may not be apparent through traditional analytical methods by focusing on the intrinsic relationships between data points. This ability to discern the underlying data structure is crucial in various AI applications, such as computer vision, natural language processing, and network analysis.

Geometry, however, studies shapes, sizes, and relative positions of objects in space. In the context of AI, geometry plays a vital role in understanding and representing the physical world. Geometric transformations, such as rotations, translations, and scaling, enable AI algorithms to manipulate and analyze data more intuitively and efficiently. This is particularly important in fields like robotics, where the accurate representation and manipulation of spatial information are critical for tasks such as navigation, object recognition, and manipulation.

Topology and geometry provide a powerful framework for AI researchers to model, analyze, and interpret complex data. By leveraging these mathematical concepts, AI algorithms can better understand the world around them, leading to more accurate and robust solutions to various problems. In the following sections, we will explore the foundational concepts of topological spaces and geometric transformations, delve into the fascinating world of manifold learning, and examine the various applications of topology and geometry in AI. Finally, we will discuss the prospects of these mathematical tools in the ongoing quest to develop more advanced and human-like artificial intelligence.

Exploring Topological Spaces: The Foundation of Geometric
Analysis

In artificial intelligence, the study of topology and geometry plays a
crucial role in understanding and interpreting complex data structures.
To appreciate the significance of these mathematical disciplines, we
must first delve into the concept of topological spaces, which serve as
the foundation for geometric analysis.

Topology, often referred to as "rubber sheet geometry," is a branch
of mathematics that deals with the properties of space that are
preserved under continuous transformations. Topology is more
concerned with the qualitative aspects of shapes and spaces than quan-
titative aspects, such as distance and angle measurements. This unique
perspective allows us to analyze and classify data flexibly and intuitively.

A topological space is a set of points and a collection of open
subsets that satisfy certain axioms. These axioms are designed to
capture the essence of continuity and closeness, which are fundamental concepts in the study of geometry. By defining a topological
space, we can explore various properties of the space, such as connectedness, compactness, and convergence, which are essential in understanding the underlying structure of the data.

One of the most important aspects of topological spaces is the
concept of a continuous function. A function between two topological
spaces is continuous if the preimage of every open set in the target
space is an open set in the domain space. This notion of continuity is
crucial in AI, as it allows us to study the behavior of algorithms and
models under small perturbations of the input data.

As we venture further into topology, we encounter the fascinating
subject of geometric transformations. These transformations, which
include translations, rotations, and scaling, are the key to unraveling
complex patterns and structures hidden within the data. By understanding how these transformations affect the topological properties of
a space, we can develop powerful tools and techniques for AI

applications, such as pattern recognition, image processing, and data visualization.

In summary, exploring topological spaces lays the groundwork for studying geometry in artificial intelligence. By understanding the fundamental concepts of continuity, connectedness, and convergence, we can develop a deeper appreciation for the hidden structures that govern the behavior of complex data sets. This knowledge, in turn, enables us to create more robust and efficient AI algorithms, pushing the boundaries of what is possible in the ever-evolving field of artificial intelligence.

Delving into Geometric Transformations: The Key to Unraveling Complex Patterns

As we venture deeper into topology and geometry, it becomes increasingly evident that geometric transformations are pivotal in uncovering the hidden structures within complex patterns. In this section, we will delve into the intricacies of geometric transformations, elucidating their significance in the context of artificial intelligence.

Geometric transformations can be defined as operations that alter an object's position, orientation, or size while preserving its essential properties. These transformations are instrumental in data analysis and manipulation, particularly in AI, where the ability to recognize and adapt to patterns is of paramount importance. There are four primary types of geometric transformations: translation, rotation, scaling, and reflection.

Translation refers to moving an object from one location to another without altering its orientation or size. In AI, this transformation is crucial for tasks such as object recognition and tracking, where the position of an object in an image or a sequence of images must be determined.

Rotation, on the other hand, involves pivoting an object around a fixed point or axis. This transformation is particularly relevant in computer vision, where the orientation of an object must be discerned to facilitate tasks such as image registration and 3D reconstruction.

Scaling entails resizing an object while maintaining its proportions. This transformation is vital in AI applications that require comparing objects with varying sizes, such as facial recognition and object detection. By scaling images or features to a standard size, AI algorithms can analyze and compare them more effectively.

Lastly, reflection involves flipping an object across a line or plane, creating a mirror image. This transformation is useful in AI applications that necessitate identifying symmetrical patterns or generating novel perspectives on existing data.

The power of geometric transformations lies in their ability to reveal hidden structures within complex patterns. By applying these transformations to high-dimensional data, AI algorithms can uncover relationships and patterns that may not be readily apparent. For instance, manifold learning techniques, such as t-distributed stochastic neighbor embedding (t-SNE) and Isomap, leverage geometric transformations to reduce the dimensionality of data, thereby facilitating the visualization and analysis of intricate structures.

In conclusion, geometric transformations are vital in the AI toolbox, enabling the discovery and manipulation of hidden structures within complex patterns. As we continue to push the boundaries of AI research and development, the significance of topology and geometry, mainly geometric transformations, will only grow. By harnessing the power of these transformations, we can unlock new insights and capabilities, propelling AI to new heights.

Manifold Learning: Unveiling the Hidden Structures in High-Dimensional Data

Manifold learning emerges as a powerful technique in artificial intelligence that allows us to uncover hidden structures within high-dimensional data. As we delve into this fascinating topic, we will explore the concept of manifolds, the significance of manifold learning, and the various algorithms employed to achieve this feat.

The Concept of Manifolds: A Gateway to Simplification

A manifold is a mathematical construct representing a continuous, smooth surface embedded within a higher-dimensional space. It is a lower-dimensional representation of complex data that retains the essential structure and relationships between data points. The concept of manifolds is crucial in manifold learning, as it enables us to simplify high-dimensional data while preserving its inherent properties.

The Significance of Manifold Learning in AI

High-dimensional data is common in AI, with applications ranging from image recognition to natural language processing. However, analyzing and interpreting such data can be daunting due to the "curse of dimensionality," which refers to the exponential increase in complexity as the number of dimensions grows.

Manifold learning comes to the rescue by reducing the dimensionality of the data, making it more manageable and easier to analyze. Manifold learning allows AI algorithms to identify patterns, make predictions, and ultimately improve their performance by unveiling the hidden structures within the data.

Algorithms for Manifold Learning: A Comparative Overview

Several manifold learning algorithms have been developed to tackle the challenges posed by high-dimensional data. Some of the most popular ones include:

- **Principal Component Analysis (PCA):** PCA is a linear technique that projects high-dimensional data onto a lower-dimensional subspace, maximizing the variance of the projected data. While PCA effectively reduces dimensionality, it may not capture the underlying non-linear structures in the data.
- **Isomap:** Isomap is a non-linear technique that seeks to preserve the geodesic distances between data points while reducing dimensionality. By constructing a neighborhood

graph and computing shortest paths, Isomap can uncover the intrinsic geometry of the data.

- **Locally Linear Embedding (LLE):** LLE is another non-linear technique that aims to preserve local relationships between data points. LLE can effectively map high-dimensional data onto a lower-dimensional manifold by reconstructing each **data point as a linear combination of its neighbors.**

- **t-Distributed Stochastic Neighbor Embedding (t-SNE):** t-SNE is a probabilistic technique that minimizes the divergence between probability distributions in high-dimensional and low-dimensional spaces. This approach is particularly effective in visualizing high-dimensional data, as it preserves local and global structures.

Manifold Learning in Action: Real-World Applications

Manifold learning has found numerous applications in AI, some of which include:

- **Computer Vision:** By reducing the dimensionality of image data, manifold learning can improve the performance of image recognition and classification algorithms.

- **Natural Language Processing:** Manifold learning can analyze high-dimensional text data, enabling AI algorithms to understand better and process human language.

- **Robotics:** In robotics, manifold learning can simplify sensor data, allowing robots to navigate and interact with their environment more effectively.

In conclusion, manifold learning is a powerful tool in the AI tool-box, enabling us to unveil the hidden structures within high-dimensional data and, ultimately, enhance the performance of AI algorithms. As we continue to push the boundaries of AI research and development, the importance of topology and geometry and manifold learning, in particular, will only grow.

Applications of Topology and Geometry in AI: From Computer Vision to Robotics

The fascinating world of topology and geometry has far-reaching implications in artificial intelligence (AI). These mathematical disciplines have become indispensable tools for researchers and developers by uncovering hidden structures and patterns in complex data. In this section, we will delve into the myriad applications of topology and geometry in AI, focusing on two key areas: computer vision and robotics.

Computer Vision: Seeing the World Through Geometric Lenses

Computer vision, a subfield of AI, aims to teach machines how to interpret and understand the visual world. Topology and geometry are crucial in this endeavor, as they provide the mathematical framework to analyze and process images and videos.

One of the most prominent applications of geometry in computer vision is feature extraction. By identifying and describing distinctive geometric structures in images, such as corners, edges, and contours, algorithms can efficiently recognize and classify objects. For instance, the Scale-Invariant Feature Transform (SIFT) and Oriented FAST and Rotated BRIEF (ORB) algorithms rely on geometric transformations to detect and describe local features invariant to scale, rotation, and illumination changes.

Topology also contributes significantly to computer vision, particularly in image segmentation. Topological data analysis (TDA) techniques, such as persistent homology, enable the identification of meaningful structures and patterns in images. By capturing the underlying topology of an image, TDA can distinguish between different objects and regions, even in the presence of noise and occlusions.

Robotics: Navigating the World with Geometric Precision

Robotics, another subfield of AI, deals with the design, construction, and operation of robots. Topology and geometry are integral to developing advanced robotic systems, as they provide the mathematical tools to model and navigate complex environments.

In robotic path planning, for instance, geometric algorithms are employed to find the shortest and safest routes for robots to traverse. By representing the environment as a geometric space, researchers can devise efficient algorithms to avoid obstacles and reach target destinations. Topological methods, such as homotopy classes, can also classify different paths based on their qualitative properties, enabling robots to choose the most suitable route according to specific criteria.

Moreover, topology and geometry are essential in developing robotic manipulation and grasping techniques. By modeling the shape and structure of objects, robots can determine the most effective way to grasp and manipulate them. Geometric reasoning can also be applied to analyze the kinematics and dynamics of robotic systems, allowing for precise control and coordination of their movements.

In conclusion, the applications of topology and geometry in AI are vast and varied, spanning from computer vision to robotics. By leveraging these mathematical disciplines, researchers and developers can uncover hidden structures and patterns in complex data, paving the way for more advanced and intelligent systems. As AI continues to evolve, the significance of topology and geometry will only grow, shaping the future of AI research and development.

The Future of Topology and Geometry in AI Research and Development

As we have journeyed through the fascinating world of topology and geometry, it is evident that these mathematical concepts play a crucial role in developing and advancing artificial intelligence. From the foundations of topological spaces to the intricacies of manifold learning, applying these ideas has enabled AI researchers to uncover hidden structures and patterns in complex, high-dimensional data. In this concluding section, we will explore the potential future directions of topology and geometry in AI research and development and the challenges and opportunities that lie ahead.

One promising avenue for the future of topology and geometry in AI is the integration of these concepts with deep learning techniques. Deep learning, a subset of machine learning, has recently gained significant attention due to its ability to learn and extract features from raw data automatically. By incorporating topological and geometric principles into deep learning algorithms, researchers can develop more robust and efficient models to understand better and interpret complex data structures.

Another exciting direction for topology and geometry in AI is the development of new algorithms and techniques that can efficiently handle large-scale, high-dimensional data. As the amount of data generated by various sources continues to grow exponentially, it becomes increasingly important for AI systems to process and analyze this information effectively. By leveraging the power of topology and geometry, researchers can create novel methods for dimensionality reduction, data compression, and pattern recognition that can scale to meet the demands of the ever-growing data landscape.

Moreover, applying topology and geometry in AI is not limited to computer vision and robotics. These mathematical concepts can also be applied to other domains, such as natural language processing, bioinformatics, and social network analysis, to name a few. By extending the reach of topology and geometry to these areas, researchers can uncover new insights and develop innovative solutions to a wide range of problems.

Despite the immense potential of topology and geometry in AI, some challenges must be addressed. One such challenge is the need for more efficient computational methods and tools to handle the complex calculations involved in topological and geometric analysis. Additionally, there is a need for more interdisciplinary collaboration between mathematicians, computer scientists, and AI researchers to further advance the understanding and application of these concepts in AI.

In conclusion, the future of topology and geometry in AI research and development is undoubtedly bright and full of possibilities. By continuing to explore and harness the power of these mathematical concepts, AI researchers can unlock new levels of understanding and innovation, ultimately pushing the boundaries of what artificial intelligence can achieve. As we move forward into this exciting future, researchers, practitioners, and enthusiasts must remain curious, open-minded, and collaborative in their pursuit of knowledge and discovery.

Chapter Summary

- Topology and geometry are essential tools in AI research, as they help uncover hidden structures and patterns in complex data, leading to more accurate and robust solutions in various applications such as computer vision, natural language processing, and network analysis.
- Topological spaces, which focus on the intrinsic relationships between data points, serve as the foundation for geometric analysis and are crucial for understanding the underlying structure of data.
- Geometric transformations, including translations, rotations, and scaling, enable AI algorithms to manipulate and analyze data more intuitively and efficiently, particularly in robotics.
- Manifold learning is a powerful technique that uncovers hidden structures within high-dimensional data, simplifying the data while preserving its inherent properties and enhancing the performance of AI algorithms.
- Popular manifold learning algorithms include Principal Component Analysis (PCA), Isomap, Locally Linear Embedding (LLE), and t-Distributed Stochastic Neighbor Embedding (t-SNE).

- Topology and geometry have numerous applications in AI, such as feature extraction, image segmentation in computer vision, path planning, and robotic manipulation in robotics.
- Future directions for topology and geometry in AI include integrating these concepts with deep learning techniques, developing new algorithms for handling large-scale data, and extending their applications to other domains like natural language processing and bioinformatics.
- Challenges in the field include more efficient computational methods, topological and geometric analysis tools, and increased interdisciplinary collaboration between mathematicians, computer scientists, and AI researchers.

THE FUTURE OF MATHEMATICS IN AI

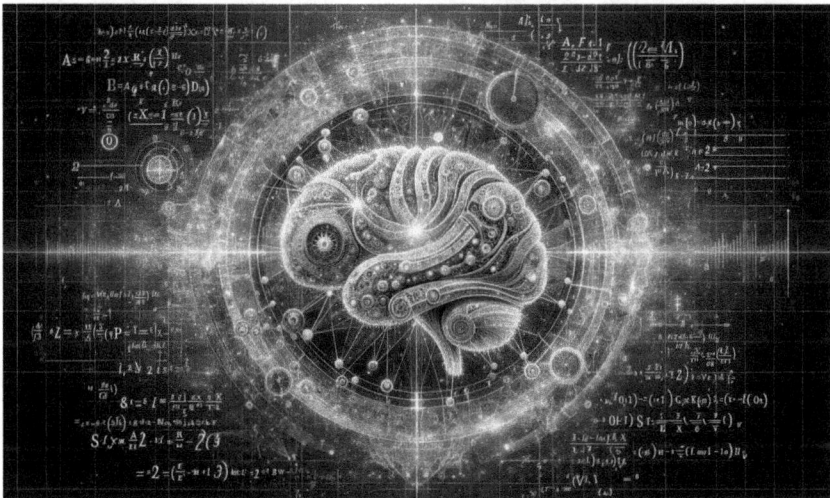

A s we stand on the precipice of a new era in artificial intelligence, it is crucial to recognize mathematics's indispensable role in shaping this rapidly evolving landscape. The purpose of this concluding chapter is to provide a comprehensive overview of the essential mathematical concepts and techniques discussed throughout this book and to explore their implications for the future of AI. By synthesizing the significant themes and findings, we aim to highlight the significance of these mathematical foundations in driving advancements in AI while also acknowledging the limitations and critiques accompanying such a complex and interdisciplinary field.

In this epilogue, we will delve into the broader implications of our findings, examining how the mathematical principles we have explored can be applied to real-world AI applications and challenges. Furthermore, we will address our approach's potential shortcomings and critiques, fostering a balanced and nuanced understanding of the subject matter. Finally, we will offer recommendations for future research and development in AI, charting a path forward that is both innovative and grounded in the rich mathematical heritage that has brought us to this point.

By providing a straightforward and engaging narrative, we aim to inspire readers to appreciate the beauty and power of mathematics in AI and recognize its potential to shape the future of technology and society.

A Comprehensive Recap

As we reach the culmination of our exploration into the world of essential mathematics for artificial intelligence, it is crucial to take a moment to reflect on the major themes and findings that have emerged throughout this journey. By revisiting these key concepts, we can better understand the significance of mathematics in the development and advancement of AI and its potential to shape the future of this rapidly evolving field.

First and foremost, we have delved into the foundational mathematical concepts that underpin AI, including linear algebra, probability and statistics, calculus, and optimization. These core principles are the building blocks for AI algorithms and models, enabling them to learn, adapt, and make data-based decisions. By mastering these essential mathematical tools, AI practitioners can harness the power of AI to solve complex problems and drive innovation across various domains.

Another central theme throughout this book is the importance of interdisciplinary collaboration. As AI continues to permeate diverse fields such as healthcare, finance, and transportation, it becomes increasingly vital for AI practitioners to work closely with domain experts to ensure that mathematical models are grounded in real-world contexts and address relevant challenges. This collaborative approach enhances the efficacy of AI solutions and fosters a deeper understanding of the underlying mathematical principles.

We have also explored the role of ethics and responsibility in AI development, emphasizing the need for transparency, fairness, and accountability in mathematical models. As AI systems become more prevalent and influential in our daily lives, we must consider the potential consequences of their actions and strive to mitigate any negative impacts. By incorporating ethical considerations into the mathematical foundations of AI, we can work towards creating a more just and equitable future for all.

Throughout this book, we have examined various case studies and real-world applications of AI, showcasing the versatility and potential of mathematics in addressing diverse challenges. From natural language processing to autonomous vehicles, these examples have demonstrated the power of mathematical models in driving AI innovation and transforming industries.

Finally, we have discussed the importance of continuous learning and adaptation in AI. As new mathematical techniques and algorithms emerge, AI practitioners must remain agile and open to change, embracing the evolving landscape of AI and its potential to revolutionize how we live, work, and interact with the world around us.

In conclusion, this comprehensive recap serves as a testament to mathematics's critical role in the development and advancement of AI. By understanding and mastering these essential mathematical concepts, we can unlock the full potential of AI and shape a future that is both innovative and responsible.

Shaping the AI Landscape

As we delve into the implications and significance of essential mathematics in artificial intelligence, it is crucial to recognize the profound impact these mathematical concepts have on the development and advancement of AI technologies. The symbiotic relationship between mathematics and AI has not only shaped the AI landscape but has also paved the way for groundbreaking innovations that continue to revolutionize various industries and aspects of human life.

One of the most significant implications of essential mathematics in AI is its role in developing machine learning algorithms. These algorithms, the backbone of AI systems, rely heavily on mathematical concepts such as linear algebra, probability, and calculus to analyze and process vast amounts of data. By understanding and applying these mathematical principles, AI researchers and engineers can create more efficient and accurate algorithms to learn from data, adapt to new information, and ultimately make intelligent decisions.

Moreover, the significance of essential mathematics in AI extends beyond machine learning. It also plays a vital role in optimizing AI systems, ensuring they operate at peak performance. For instance, mathematical concepts such as graph theory and combinatorics are instrumental in solving complex optimization problems in AI applications, such as routing and scheduling in logistics or resource allocation in cloud computing.

Furthermore, essential mathematics is a foundation for developing new AI techniques and methodologies. As AI evolves, researchers constantly explore novel mathematical concepts and theories that can unlock new capabilities and applications for AI systems. For example, the emerging field of topological data analysis, which leverages the principles of algebraic topology, has shown promise in enhancing the robustness and interpretability of machine learning models.

In addition to its direct impact on AI technologies, essential mathematics fosters interdisciplinary collaboration and innovation. By bridging the gap between mathematics and AI, researchers from diverse fields such as computer science, engineering, and even social sciences can come together to tackle complex problems and develop innovative solutions that have far-reaching implications for society.

However, the significance of essential mathematics in AI has its challenges. As AI systems become increasingly sophisticated and complex, the demand for individuals with a strong foundation in mathematics and AI-related disciplines continues to grow. To ensure that the AI landscape continues to thrive and evolve, it is imperative to invest in education and training programs that equip the next generation of AI researchers and engineers with the necessary mathematical skills and knowledge.

In conclusion, the implications and significance of essential mathematics in AI are vast and far-reaching. By understanding and harnessing the power of these mathematical concepts, we can continue to shape the AI landscape, drive innovation, and unlock the full potential of artificial intelligence in transforming our world for the better.

Addressing the Inevitable Shortcomings

As we delve into the limitations and critiques of essential mathematics in AI, it is crucial to acknowledge that no field is without its shortcomings. This section will address some of the most pressing concerns and challenges when applying mathematical concepts to artificial intelligence. By doing so, we aim to provide a balanced perspective and encourage further research and development in this fascinating domain.

The Complexity Conundrum

One of the most significant limitations of using mathematics in AI is the inherent complexity of both fields. As AI systems become more advanced, the mathematical models and algorithms required to support them grow increasingly intricate. This complexity can make it difficult for researchers and practitioners to fully comprehend and optimize the AI systems they are working with. Moreover, the steep learning curve associated with mastering advanced mathematical concepts may deter some individuals from pursuing careers in AI, potentially limiting the field's growth and innovation.

The Black Box Dilemma

Another critique of using mathematics in AI is the so-called "black box" problem. This issue arises when AI systems become so complex that their inner workings are difficult, if possible, to understand. As a result, explaining how these systems arrive at their conclusions can take time and effort, leading to concerns about transparency, accountability, and trustworthiness. While mathematical models can provide a solid foundation for AI systems, they may only sometimes offer clear insights into the decision-making processes of these systems.

The Bias Pitfall

The application of mathematics in AI is not immune to the issue of bias. AI systems are often trained on large datasets, which may contain inherent biases that can be inadvertently perpetuated by the algorithms. Furthermore, the mathematical models may be designed with certain assumptions that introduce bias into the AI system. Addressing these biases is essential to ensure that AI systems are fair, ethical, and effective.

The Adaptability Challenge

Lastly, one of the limitations of using mathematics in AI is the difficulty in adapting mathematical models to the ever-evolving landscape of AI. As new technologies and techniques emerge, keeping mathematical models up-to-date and relevant can be challenging. This adaptability challenge underscores the importance of interdisciplinary collaboration between mathematicians, computer scientists, and other experts in the field of AI.

In conclusion, while essential mathematics plays a critical role in the development and advancement of AI, it is important to recognize and address the limitations and critiques accompanying its application. By doing so, we can work towards refining our understanding of the complex relationship between mathematics and AI, ultimately contributing to the growth and success of this exciting field.

Charting the Path Forward

As we reach the end of our mathematical journey through artificial intelligence, we must reflect on the insights gained and the path ahead. Mathematics has always been the backbone of scientific progress, and its role in AI is no exception. In this final section, we will provide a synthesis of our findings and offer recommendations for future research and development in the field of AI.

Throughout this book, we have explored the essential mathematical concepts that underpin AI, delving into topics such as linear algebra, probability theory, optimization, and more. We have seen how these mathematical tools have enabled the development of sophisticated AI algorithms and models, empowering machines to learn, reason, and adapt in ways that were once the exclusive domain of human intelligence.

The implications of these advancements are profound, with AI poised to revolutionize industries, economies, and societies across the globe. From healthcare and education to finance and transportation, the potential applications of AI are vast and varied. However, as we have also discussed, some limitations and critiques must be acknowledged and addressed to ensure the responsible and ethical development of AI technologies.

In light of these considerations, we offer the following recommendations for charting the path forward in the field of AI:

- **Foster interdisciplinary collaboration:** The development of AI requires expertise from a diverse range of fields, including mathematics, computer science, engineering, and the social sciences. Encouraging collaboration between these disciplines will facilitate the exchange of ideas and promote the development of innovative solutions to complex AI challenges.

- **Prioritize ethical considerations:** As AI continues to permeate various aspects of our lives, it is essential to ensure that ethical considerations are at the forefront of AI research and development. This includes addressing issues such as algorithmic bias, data privacy, and the potential displacement of human labor.
- **Invest in education and training:** To keep pace with the rapid advancements in AI, it is crucial to invest in education and training programs that equip individuals with the necessary mathematical and computational skills. This will help address the growing demand for AI talent and empower individuals to participate in the AI-driven economy.
- **Promote open research and collaboration:** The sharing of knowledge and resources is vital to the continued progress of AI. By promoting open research and collaboration, we can accelerate the pace of innovation and ensure that the benefits of AI are accessible to all.
- **Encourage public engagement and dialogue:** As AI technologies become increasingly integrated into our daily lives, it is essential to foster public engagement and dialogue around the potential benefits, risks, and ethical implications of AI. This will help to ensure that AI development is guided by the needs and values of the communities it serves.

In conclusion, the future of mathematics in AI is both promising and challenging. By embracing the recommendations outlined above, we can work together to harness the power of mathematics and AI for the betterment of society while also addressing the potential pitfalls that may arise along the way. The journey may be complex, but the rewards are undoubtedly worthwhile.

ABOUT THE AUTHOR

Andrew Hinton is a prolific author specializing in Artificial Intelligence (AI). With a background in computer science and a passion for making complex concepts accessible, Andrew has dedicated his career to educating others about the rapidly evolving world of AI. His debut series, AI Fundamentals, is a comprehensive guide for those seeking to understand and apply AI in various professional settings. Andrew's work caters to a broad audience, from managers to coders, breaking down AI basics, essential math, machine learning, and generative AI clearly and engagingly. His ability to demystify the complexities of AI has made him a trusted voice in the tech industry. Andrew's work imparts knowledge and empowers his readers to navigate and innovate in an AI-driven world.

www.ingramcontent.com/pod-product-compliance
Lightning Source LLC
Chambersburg PA
CBHW050643190326
41458CB00008B/2390